A NOT-SO-NEW
WORLD

EARLY AMERICAN STUDIES
Series editors:
Daniel K. Richter, Kathleen M. Brown,
Max Cavitch, and David Waldstreicher

Exploring neglected aspects of our colonial,
revolutionary, and early national history and culture,
Early American Studies reinterprets familiar themes
and events in fresh ways. Interdisciplinary in character,
and with a special emphasis on the period from about
1600 to 1850, the series is published in partnership with
the McNeil Center for Early American Studies.

A complete list of books in the series
is available from the publisher.

A NOT-SO-NEW WORLD

Empire and Environment
in French Colonial North America

Christopher M. Parsons

PENN

UNIVERSITY OF PENNSYLVANIA PRESS

PHILADELPHIA

Published by
University of Pennsylvania Press
Philadelphia, Pennsylvania 19104-4112
www.upenn.edu/pennpress

Printed in the United States of America on acid-free paper
1 3 5 7 9 10 8 6 4 2

Library of Congress Cataloging-in-Publication Data
Names: Parsons, Christopher M., author.
Title: A not-so-new world : empire and environment in
    French colonial North America / Christopher M. Parsons.
Other titles: Early American studies.
Description: 1st edition. | Philadelphia : University of
    Pennsylvania Press, [2018] | Series: Early American studies |
    Includes bibliographical references and index.
Identifiers: LCCN 2018004263 | ISBN 9780812250589
    (hardcover : alk. paper)
Subjects: LCSH: Canada—History—To 1763 (New France) |
    France—Colonies—America—History. | North America—
    Environmental conditions—History. | Imperialism—
    Environmental aspects. | Horticulture—North America—
    Foreign influences—History. | Imperialism and science—
    France—History.
Classification: LCC F1030 .P268 2018 | DDC 971.01—dc23
LC record available at https://lccn.loc.gov/2018004263

CONTENTS

The View from Champlain's Gardens

In the fall of 1608, Samuel de Champlain began to cultivate a New France at Québec. "On the first of October I had some wheat sown and on the fifteenth some rye," he wrote, before continuing and explaining that "on the third of the month there was a white frost and on the fifteenth the leaves of the trees began to fall. On the twenty-fourth of the month I had some native vines planted and they prospered extremely well."[1] That he focused his account on his efforts to sow grains and transplant vines is not surprising; these were symbolically laden plants that occupied a central place in France's sense of its own past and its colonial future. With roots in both biblical history and the classical colonialism of Rome, the presence of these plants provided both evidence of France's moral authority to claim indigenous lands and an augur for the ultimate success of colonialism on a continent that had thus far proven resistant to French ambitions.[2] As the horticulturalist Olivier de Serres had recently written in his innovative and influential *Théâtre d'agriculture et mesnage des champs*, "After bread comes wine, second food given by the creator for the maintenance of life, & the first celebrated for its excellence."[3] Although unified in symbolism, the act of planting grains and grapes was also the staging of an encounter between introduced European seeds and indigenous American vines that Champlain had had transplanted in the colony's gardens. Orchestrating this encounter and celebrating this proximity became a founding act that set French roots into a new not-so-New World.

In the preceding five years, Champlain (and French colonial efforts more generally) had focused on settlements in Acadia. At St. Croix Island (1604) and Port Royal (1605) in what are now, respectively, Maine and Nova Scotia, Champlain had similarly laid out gardens as he studied landscapes and prepared for shelter and defense. When the settlement that he called the *habitation* was established at Québec on July 3, Champlain continued these practices. He began to cultivate his new settlement during a challenging summer and fall. He had only recently survived an attempted mutiny, and bitter experience in recent winters had impressed the urgent need to prepare supplies and support for harsh winters.[4] It is likely, as well, that more than concerns about survival weighed on Champlain as Québec took shape in front of him. Only the year before, he had returned to France with the settlers of Port Royal as the crown withdrew its support for the Acadian colony and demanded that its leaders justify further investments in the settlement of French colonies in the region.[5] For both Champlain and the broader colonial effort in which he was an integral part, the stakes were high as the first frost fell and winter threatened its arrival.

Gardens were an essential element of the expanding footprint of French colonialism in northeastern North America. "While the carpenters, sawyers and other laborers worked at our lodgings," Champlain wrote, "I put all the rest to clearing around the *habitation*, so as to make gardens in which to sow seeds to see how they would all succeed, for the soil appeared quite good."[6] The grape-bearing vines that the explorer had transplanted into colonial gardens took pride of place in the written account of these first months at Québec. When, the following year, he returned to France to inform his superiors of his progress, the vines were neglected and he complained bitterly about their loss. He explained that "after I had left the *habitation* to return to France, they were ruined for lack of care, which distressed me very much at my return."[7] Champlain was upset again when the same thing happened in 1610–11. Those who had been left in charge of the gardens "took no action to conserve them . . . [and] at my return, I found them all broken, which brought me a great displeasure, for the little care that they had had for the conservation of such a good and beautiful plot, from which I had promised myself something good would come."[8]

Other colonists also cultivated gardens as France's colonial presence pushed out from the *habitation*. Rather than the closed garden familiar to medieval Europe (the *hortus conclusus*), these too were created as ambitious projects that aimed to engage with and reshape American environments. The lawyer, colonist, and author Marc Lescarbot had earlier drawn Renaissance-inspired gardens in the landscape of early Acadia.[9] When the Récollet missionary Gabriel Sagard traveled to the region that he called Canada in 1625, he described the gardens of his order as "very beautiful, and of a good base of soil; for all of our herbs and roots do well there, and better than many

A Le magazin.
B Colombier.
C Corps de logis où font nos armes, & pour loger les ouuriers.
D Autre corps de logis pour les ouuriers.
E Cadran.
F Autre corps de logis où eſt la forge; & artiſans logés.
G Galleries tout au tour des logemens.
H Logis du ſieur de Champlain.
I La porte de l'habitation, où il y a Pont-leuis.
L Promenoir autour de l'habitation contenant 10. pieds de large iuſques ſur le bort du foſſé.
M Foſſés tout autour de l'habitation.
N Plattes formes, en façon de tenailles pour mettre le canon.
O Iardin du ſieur de Champlain.
P La cuiſine.
Q Place deuant l'habitation ſur le bort de la riuiere.
R La grande riuiere de ſainct Lorens.

Figure 1. Samuel de Champlain, "Habitation of Quebec," *Les voyages du Sieur de Champlain*, 1613. Courtesy of the John Carter Brown Library at Brown University, Providence, Rhode Island.

gardens that we have in France."[10] Like Champlain's *habitation*, these landscapes blurred the transition across a trench that surrounded the residence and
fields of "flowers, particularly those that we call *Cardinales* and *Martagons*,"
and raspberry bushes that soon became part of the colonial diet.[11] The effect
was that his order's residence resembled "a little house of the rural nobility,
rather than a monastery of the Franciscan friars."[12] Nearby on the plateau,
Sagard reported seeing the cultivated lands of the apothecary and colonist
Louis Hébert, and in particular "a young apple tree that was brought from
Normandy and young vines which were very beautiful."[13] These would soon be
joined by the gardens of the Society of Jesus after missionaries from this order
came to join the Récollets in New France, as well as those areas of cultivation
that continued to expand from the original site of settlement at Québec.[14]

     Cultivated spaces were well represented in images of Champlain's settle-
ment of Québec that show gardens touching up against the margins (Figure 1).
They echo similar features in representations of earlier settlements at Île Sainte-
Croix and Port Royal also included in Champlain's 1613 *Voyages* (Figure 2). In
each case, the prominence and location of these intricately designed landscapes amplified Champlain's focus on the cultivated spaces in his written text
and changed the tone of images that might otherwise be read as defensive or
even hostile to local environments. At both Québec and Port Royal, the *habitations'* firing cannons could have spoken to a desire to cut off contact with
the outside world to which the French had come. Yet as the expansive gardens
in the image pushed at the borders of the image, they effectively blurred the
division between the fortress-like *habitation* that occupied the center of the
composition and the natural world around it. No matter how far the smoke of
the cannons could drift from the *habitation*, Champlain's intricately laid-out
parterres extended beyond them.

     In images such as these, tilled fields and cultivated gardens became rec-
ognizable as sites in which the novelty of New France was tested. One of the
foundational acts of Champlain's settlement, after all, was the introduction
of local flora—grains and vines included—into French colonial environments.[15] Québec's gardens were experiments as much as they were a source of
sustenance, and they welcomed a wide range of American and European
plants including "kitchen herbs of all sorts with very beautiful corn, wheat,
rye, and barley that were sown and some vines that I had planted during the
wintering."[16] Some grew from seeds that had been brought across the Atlantic, but others had homes in the forest and fields that Champlain so passion-
ately described as he ascended the Saint Lawrence River.[17] Near Québec, he

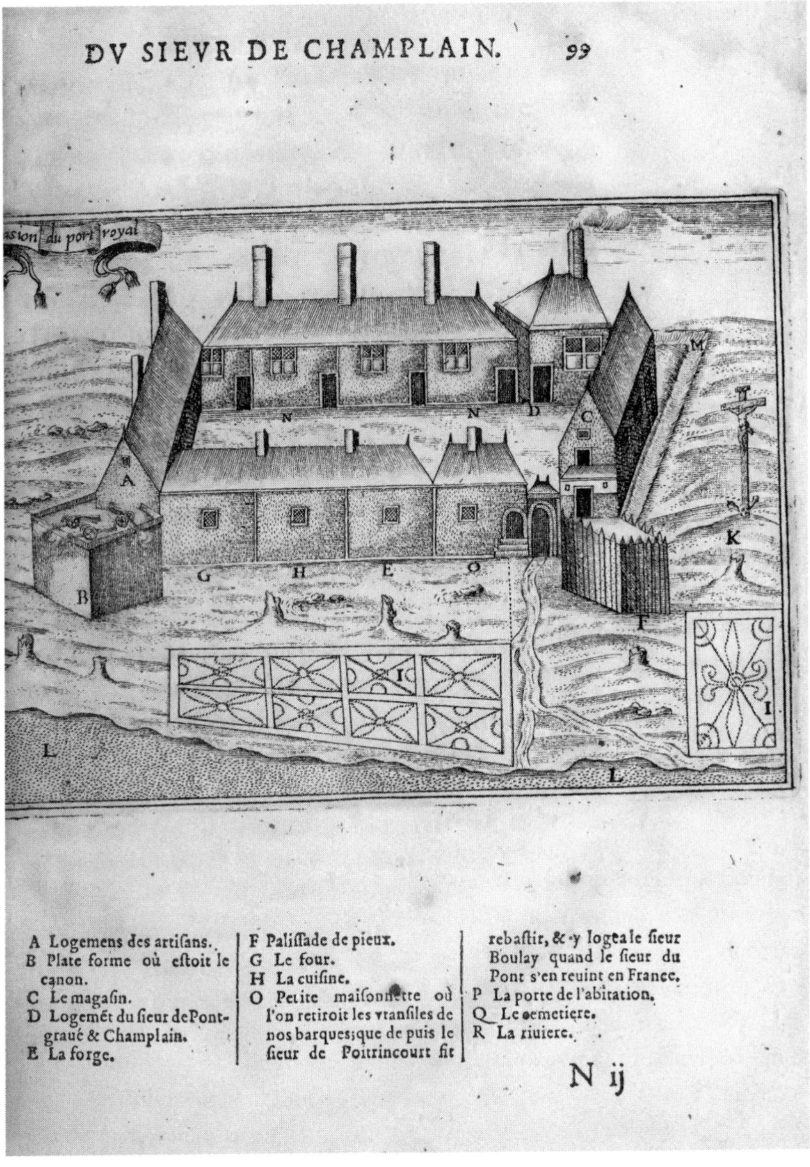

DV SIEVR DE CHAMPLAIN. 99

A Logemens des artiſans.
B Plate forme où eſtoit le canon.
C Le magaſin.
D Logemét du ſieur dePont-graué & Champlain.
E La forge.
F Paliſſade de pieux.
G Le four.
H La cuiſine.
O Petite maiſonnette où l'on retiroit les vtanſiles de nos barques; que de puis le ſieur de Poitrincourt fit rebaſtir, & y logea le ſieur Boulay quand le ſieur du Pont s'en reuint en France.
P La porte de l'abitation.
Q Le cemetiere.
R La riuiere.

N ij

Figure 2. Samuel de Champlain, "Port Royal Habitation," *Les voyages du Sieur de Champlain*, 1613. Courtesy of the John Carter Brown Library at Brown University, Providence, Rhode Island.

described seeing "all of the types of trees that we have in our forests on this side of the ocean, and a number of fruits, even though they are *sauvage* for lack of cultivation: such as butternut, cherries, plums, vines, raspberries, strawberries, and green and red currants."[18] Elsewhere he noted "vine that produces reasonably good raisins, even though it is *sauvage*, which once transplanted & labored, will produce fruits in abundance."[19] It was these plants—roughly similar as recognizable *sauvage* (wild) versions of plants that grew in Europe— that Champlain brought into his gardens. They were, or perhaps could become, "like those we have in France."[20] These were spaces that visually represented a managed abundance that naturalized French colonialism as a well-governed and desirable hybridity.[21]

When Champlain called the grapevines and American plants at Québec *sauvage* he marked them as a site of tension between the familiar and the foreign. *Sauvage* has more often been studied when it was used as a noun within French colonial discourse to refer to indigenous peoples, but the term itself blurred distinctions between the human and non-human world.[22] The word pointed less toward a neat ontological distinction than, as historian Sophie White has recently written, a "protean" marker of identity defined by its "flexibility."[23] It marked out a spatial and cultural liminality.[24] The *Thresor de la langue francoyse*, written by Jean Nicot in 1606, provided synonyms including "semiferus," "sylvestris," and "erratico." If "semiferus" suggested a partial ferality, "sylvestris" and "erratico" instead connected the term to the forest or an errant path and from there to conceptual and spatial borderlands. The entry further suggested that it was possible to "become *sauvage*" or "make or render all *sauvage*," highlighting that the state identified as *sauvage* was neither fixed nor completely conquerable.[25] Where generations of historians translated the term simply as "savage," scholars now often prefer to interpret it as "wild" or leave it untranslated altogether, as I do here.[26]

When Champlain wrote that local flora was "*sauvage* for lack of cultivation," he signaled that the manifest differences of American environments were remediable defects and hinted at the ecological ambitions of France's colonial project.[27] As a metaphor, the discourse of cultivation evoked by *sauvage* related the process of conversion to the domestication of wild plants and animals.[28] In the colonial context of northeastern North America, the domestication of place and peoples became explicitly linked by colonists who cast themselves as cultivators and framed empire itself as an act of rehabilitation. The discourse of the *sauvage* was in this way overtly ecological and blurred ethnographic and environmental knowledge. It is worth remembering that "culture," in the period discussed here, was still largely an agricultural term, familiar most often in the context of the "culture of the earth."[29] It was at this time gradually being displaced toward its modern anthropocentricity by authors who perceived, in the term's evocation of domestication, elevation, and refinement, a resonance with a new optimism in human progression and cultural evolution.[30] As a colonizing ideology, a focus on culture and prescriptions for cultivation therefore inspired practice and encouraged active engagement with the places and peoples of this not-so-New World in a bid to understand and ultimately assimilate them both.[31] In this, it had much in common with the equally generative English term "improvement."[32] From the founding moments of New France and Champlain's first efforts to turn over the soils of Québec, an emergent colonial political ecology privileged cultivation as a way of both knowing and creating a New France in North America.[33] As the Jesuit Pierre Biard explained from his mission in Acadia, northeastern North America could be remade in the image of Europe; it was, he wrote, "another France . . . to be cultivated."[34]

*　　*　　*

The impulse to cultivate a New France produced deep and lasting footprints in the human and natural histories of North America. Yet it is a different environmental history of northern North America than has often been told. The gardens of Québec might seem strange places from which to proceed, when the French colonized regions of the continent that many are more likely to associate with cold-weather animals such as cod and beaver than they are to associate them with delicate plants and verdant landscapes.[35] In spite of the work of generations of historians who have dug down into the rich archives of agricultural settlement along the Saint Lawrence River and brought a colonial project that was firmly committed to agricultural development to light, we nonetheless remain more likely to think of the region as a wintry white rather than in shades of green and brown.[36] We might ask what room there was for gardens in what one scholar has recently characterized as a "reluctant land."[37] Yet if we notice how Sagard's narrative draws out as he described his order's garden or how some of the white space on Champlain's maps was filled in with his, we must appreciate the cultural and intellectual significance of these sorts of cultivated landscapes for French colonists. Inhabiting these sites and describing them for Atlantic audiences helped these authors conceptualize what colonialism could be in North America as well as their own place in its unfolding history.

Producing these cultivated spaces foregrounded environmental encounters in New France as a central way to know and claim colonial North America. Horticulture was a political act as well as a necessity for survival. The early architects of French colonial expansion in North America possessed a worldview that, anachronistically, we might call ecological. Cultivation offered early modern French colonists a capacious language with which to describe the interrelatedness of the human and natural environments discovered there—the culture of American peoples and the culture of American places—and to propose colonialism as a process that would transform them both. Frenchness was a quality that applied to both specific peoples and places and that could be brought out with diligent attention.[38] In this, the French who came to North America possessed a worldview that strangely mirrors our own. We, in a time of ecological crisis, find ourselves confronted with phenomena such as global warming that we recognize as neither wholly natural nor entirely social and that, in effect, have forced us to come to terms with the entanglements of the non-human and human worlds and the limits of anthropocentric approaches to understanding and inhabiting the world in which we live.[39] The character of French texts in which the weather, plants, and animals are agents in the history of colonial expansion and development can nonetheless be startling. For what strikes the reader of the accounts of early colonists, missionaries, and explorers is the extent to which these actors expected to encounter the agency of American environments and how they anticipated interaction with the world around them. In truth, we cannot say that the environment was simply acted upon by colonialism or the inverse. Colonialism and colonial experience in New France emerged through complex negotiations with place.

This is the horizon that cultivation opens for us here. The French colonists who came to northeastern North America to cultivate a New France mapped a distinction between cultivated and *sauvage* onto North American places and peoples.[40] These were lands that had been shaped by millennia of indigenous cultivation and that continued to be shaped by peoples whose ecological practices had drawn out the richness of northern environments, but colonial texts effaced indigenous labor that had coaxed the expansion of temperate flora. Colonists read evolutionary and geological histories that had produced ecological affinities between the environments of New France and Old in such a way as to justify the imposition of French colonialism. They anticipated the resistance of a wild animal bucking efforts at domestication, and they forgave themselves the violence of pruning, grafting, and transplanting in advance. The language of cultivation translated these activities into the establishment of a mutually beneficial patriarchal order and colonial dispossession into the fantasy of a well-managed agricultural estate. The sites in which these exchanges (and confrontations) took place were privileged in the texts that communicated New France to Europe as they became evidence of the extent to which French colonialism differed from the violent conquests of the Spanish in Central and South America.[41] Cultivation became the self-legitimating practice of French settler colonialism, transferring sovereignty and authority to colonists and missionaries who adopted the role of cultivator and benevolent patriarch.[42] The production of knowledge in material space and through material practice encourages us to see that it too had a footprint.[43] Representing the New World changed it and all the people who claimed it for their own.

The colonists, missionaries, merchants, and administrators who came to New France were part of a broader Atlantic culture that increasingly valued empirical observation and numbered among those in the Americas who were becoming more confident in their ability to know extra-European environments.[44] The French authors who positioned themselves as cultivators paid close attention to the question of where and how knowledge was produced, and they claimed an epistemological privilege by virtue of their geographical location. They were participants in a broader valorization of empirical study of the natural world then underway throughout the Atlantic world.[45] If there is nothing in this statement that could not be (and has not been) written about colonists in other Atlantic empires, focusing on cultivation directs us to consider how intellectual revolutions in the Atlantic world produced local and regional histories in environmentally distinct regions of the Americas. As colonists such as Champlain claimed that the *sauvage* nature of American flora was neither natural nor necessary, they foregrounded experiences gathered through touch, taste, smell, and sight to promise that an essential familiarity lay behind apparent differences in American environments; this was the essential act of an agricultural alchemy that transmuted colonialism into cultivation.[46]

French colonialism effaced indigenous labor and delegitimated indigenous knowledge as its architects suggested American environments remained unfinished and even unnatural. Throughout the wider Atlantic world, the boundaries that divided nature from artifice or from the preternatural or unnatural were contested and defined in practice.[47] The question of the natural was particularly problematic in a colonial context where French authors (and authorities) looked to representations of indigenous culture and environments to justify settlement and colonialism. Producing and policing a firm distinction between the *sauvage* and the natural was, then, the central function of French colonial political ecology in this period. Cultivation provided both a method of study and a means of intervention.

*   *   *

Many scholars have assumed that the self-evident novelty of North America posed a significant challenge to European intellectual traditions that, built on Aristotelian insights into the natural world, were unable to grapple with the diversity and difference of American nature. At the very least, according to this view, the experience of American difference severely challenged the validity of classical sources that had long defined the flora and fauna of the Old World.[48] Yet while plants and animals both marvelous and monstrous may have posed a significant intellectual problem for Christopher Columbus and later explorers of Spanish territories in Central and South America, French colonists and missionaries had a fundamentally different experience of North American environments that appeared strangely familiar at first encounter. French settlers, traders, and missionaries discovered new places in the early seventeenth century that they considered novel in the way of the *Terre neuve* or Newfoundland located just to the northeast of New France. Yet novelty in the way that it would be later articulated by Georges-Louis Leclerc, the Comte de Buffon—as a chronologically new continent, manifestly different from a more familiar, knowable, and reliable Old World—was a conceptual impossibility within a worldview that insisted on the mutability of American difference.[49] It was only gradually that both colonial and metropolitan authors struggled with how to adequately capture (and contain) this American difference, opening up spaces where new forms of knowledge could be articulated and initiating a broader discussion of natural difference in French North America.

Moments of shock or marvel are rare in the accounts of seventeenth- and eighteenth-century New France. Instead, we are presented with historical actors who seem to know exactly where they are and who, because of its obviousness, can remain frustratingly reticent in their descriptions of colonial environments or in reflections on their own ecological knowledge. *A Not-So-New World* follows the rise and subsequent fall of cultivation as an organizing

ideology of French colonialism in northeastern North America as a means to
bring an otherwise obscured political ecology to light. Cultivation as a discourse translated French colonialism into a recognizable horticultural act and
shaped indigenous and colonial environments as a set of ecological and material practices. From seventeenth-century prominence to eighteenth-century
obsolescence, the story of cultivation therefore weaves together a dizzying
cast of historical actors and an equally diverse set of ecological and cultural
contexts. If it is a story of local encounters between colonists and indigenous
plants and peoples, cultivation's history nonetheless also remains entangled
with global and local ecological and climatological oscillations, intellectual
revolutions transforming the relationship with the natural and material world
in Europe, and shifting political and cultural landscapes in colonial North
America and the Atlantic world. *A Not-So-New World* therefore charts a course
that blends insights from environmental history and the history of science
to reconsider the history of European colonialism in North America, but it
follows the path laid by historical actors who self-consciously considered
their own place in its long history.

The first chapter of *A Not-So-New World* situates French exploration and
settlement in northeastern North America within broader ecological and
geological histories. It is these histories—told through archaeological remains
of plant materials and the science of plate tectonics—that can help us understand why early explorers such as Champlain seemed so at home in what we
know to be a new world for them. Shared genetic relationships between European and American flora meant that many parts of these environments were
recognizable, and we can reread narratives of exploration and settlement to see
how these familiar spaces were sought out and foregrounded in colonial accounts. These authors drew on contemporary natural science as they identified colonial flora as *sauvage* (wild) in accounts that both highlighted observable
differences between European and American flora and implied that these
were remediable defects.

The accounts in which the *sauvage* nature of New France was examined
valorized personal experience and observation and communicated the certainty that French colonialism would redeem American environments in genres
such as the travel narrative that celebrated empirical exploration. Chapter 2
examines how the texts through which New France was communicated to
French audiences made cultivation a central mode for understanding the
colony. The power of cultivation derived, in part, from a broader renaissance
in horticultural practice and theory in contemporary France. As they reported

their own experiences of colonial environments and took the measure of ecological continuities and differences, authors such as Champlain transformed
discussions of colonial environments into opportunities to theorize what
French colonialism could do in North America. Claims to know American
flora and environments translated into privileged claims to articulate the prospect and purpose of this New France. In focusing on the opportunities to
rehabilitate individual plants and whole landscapes, empire itself was recast
as a recuperative project.

Chapter 3 follows missionaries who sought to intervene in the ecological
lives of indigenous people in order to reform their spiritual lives. Looking to
missionary experience in other parts of the Americas, Jesuits and Récollets organized their initial efforts into New France around a plan to sedentarize
Algonquian peoples whom they understood to be errant and wild. At sites such
as Sillery, missionaries who presumed that indigenous people lacked ecological knowledge instead came to understand its complexity and to appreciate
the inability of their own practices to support communities in the boreal forests
of the Canadian Shield. As they moved west into Wendat territories, Jesuit
missionaries similarly discovered that mission strategies founded on the belief that indigenous peoples lacked a complete ecological knowledge excluded
them—as unmarried men hostile to many of the traditional institutions of
Iroquoian communities—from the sites of cultivation in these communities.
An imperative to cultivate indigenous peoples therefore introduced missionaries to new perspectives on indigenous places and hinted at the limits of
French ability to intervene in the lives of American places and peoples.

As cultivation encouraged empirical observation and the conscious assessment of experience, it therefore also created the conditions in which its limitations became evident. Chapter 4 examines how, after the optimistic belief
that American plants were *sauvage* versions of those in France was met with a
century of failed experiments, the imagined footprint of French colonialism
began to retreat. French colonists relied more thoroughly on French crops and
began to identify an essential foreignness in those that they encountered outside of colonial settlements. They also engaged in debates about where French
colonialism could best take root that were far more pessimistic about the ability
of colonialism to reshape ecosystems and climates. Was the Saint Lawrence
Valley simply too cold, or could the region that was increasingly called Canada
be scrubbed free of American resistance and made hospitable for French
ecological regimes? On both sides of the Atlantic, proponents of colonialism
sought to explain harsher-than-expected climates, failed introductions of

European crops, and recalcitrant American flora that refused to become identical to its French counterparts. The result was a considerable debate that foregrounded studies of North American environments as a key site for the articulation and contestation of imperial ideologies.

By the eighteenth century, disputes between French colonists and newly confident and organized metropolitan scientists about how to know North American environments mapped closely onto broader currents in the Atlantic world that saw epistemological authority centralized in major European cities. Chapter 5 introduces an institution that had a major influence on how knowledge and plants circulated in the French Atlantic world in the eighteenth century: the Paris-based Académie Royale des Sciences. The Académie was instrumental in the effort by Louis XIV and his influential minister Jean-Baptiste Colbert to centralize cultural production and authority in France, and its history is often examined as a facet of the transition to a bureaucratic state in seventeenth- and eighteenth-century Europe. The Académie became a global institution because of the ability of its American correspondents to translate the priorities—and the wages—of the Parisian institution to meet the expectations of the colonial and indigenous people who most often actually performed the act of collecting and guiding. The Académie was remarkably successful at mobilizing administrative and military networks to scientific ends, but it also drew on and manipulated a vibrant conversation between diverse colonial and indigenous populations that had begun with the first tentative efforts to establish a French colonial presence. The success of its science—a science of novelty—further marginalized the calls to cultivate a New France that had animated colonists, missionaries, and explorers in the preceding century.

The final chapter of this book examines an episode where colonial, metropolitan, and indigenous ways of studying American flora met and conflicted over the question of whether New France was essentially familiar or an entirely new and foreign continent: the multiple discoveries of American ginseng. When Joseph-François Lafitau, a Jesuit missionary to the Haudenosaunee (Iroquois), claimed to have discovered ginseng south of Montréal, he also announced his reliance on indigenous peoples. In fact, he used his discovery to claim that the existence of an Asian plant in North America proved larger ecological and cultural continuities between the Old World and the New. His arguments were disputed by French naturalists, who dismissed his ethnographically inclined method and his larger claims of global cultural and ecological continuities; theirs was a science of difference and novelty. Ultimately, when Lafitau's larger arguments were eventually accepted, it was merchants who acted fastest and who organized large-scale trade with China in the 1730s and 1740s. The trade proved disastrous for indigenous ecologies in North America. Lafitau's pursuit of physical proof of the Old World origins of indigenous cultures almost drove the plant to extinction and threatened a real botanical relationship between Eurasia and the Americas.

I am ultimately less interested in studying the accuracy of French claims to know New France than in examining the long negotiation between French discourse and American matter through which this knowledge was produced.[50] Whether in Québec or Paris, knowing New France meant engaging with a multicultural, epistemologically diverse Atlantic world. The obvious failure of efforts to weed out the features that made the environments of New France different from those that colonists had left behind also created new intellectual and cultural spaces in which indigenous knowledge was interrogated and integrated. Ultimately, French knowledge was always produced in dialogue with indigenous knowledge, whether colonial accounts explicitly praised or critiqued indigenous practice and even if they rendered aboriginal cultures invisible in celebrating the heroism of borderland scientists.[51]

*   *   *

Decades before the Comte de Buffon and Thomas Jefferson took up the infamous "Dispute of the New World," authors on both sides of the French Atlantic world were engaged in a vigorous debate about Americanness.[52] Colonist-authors, administrators, scientists, and missionaries studied and hypothesized the depth and cause of this increasingly undeniable difference. Was New France more new or more French? The simple repetition of the appellation "New France" in historical accounts and recent scholarship obscures the fact that it was the tension between these two terms that animated colonial and transatlantic debates about the nature of empire in the French North Atlantic during the seventeenth and eighteenth centuries. Throughout the century and a half of French colonialism in North America, investigations of the environments of New France were integral to these broader considerations, providing a space for colonial and metropolitan methods to assay otherwise ineffable questions of cultural transfer from one continent to another, of the conversion of novel places and peoples, and, in effect, of France's ability to extend and replicate itself beyond its shores.

CHAPTER I

Discovering a Not-So-New World

In 1623, the Récollet missionary Gabriel Sagard's first glimpse of the nascent New France to which he had come was a garden. At the mouth of the Saint Lawrence, near Mont-Sainte-Anne in Gaspé Bay, he encountered a landscape that he described as "very mountainous and elevated almost everywhere, disagreeable and sterile." He was disappointed to see "nothing but fir trees, birch, and little other wood."[1] These were the lands of the Canadian Shield, scraped bare of much of their soil by millennia of glacial movement and home to boreal forests.[2] Yet he was soon to find that "in front of the harbor, in a slightly elevated place, a garden was made that the sailors cultivated when they arrived there, they sowed sorrel, and other little herbs there, with which they make soup."[3] Amid seemingly endless forests, in a frequently harsh and forbidding landscape, he had found a garden; it was a *locus amoenus* in a New World otherwise unknown to him.

Sagard's encounter with the garden changed the tenor of his descriptions and marked a new relationship between the author and the environments he now set about to explore. As he passed into the Saint Lawrence, the Récollet encountered far more familiar and pleasing landscapes, with recognizable geology, weather, and flora.[4] Rough similarities in rainfall, in seasonality, and in ecological organization make western Europe and eastern North America the heart of a region identified by environmental scientists as the temperate forest biome.[5] Sagard would have seen large swaths of deciduous forests but with prominent stands of coniferous trees throughout.[6] The unique climates of the Saint Lawrence Valley made the region home to multiple plant communities that blended boreal elements with analogues in Scandinavia and the Baltic and genuinely temperate plants that were close relatives of those found throughout Atlantic France. Populations of familiar plants were particularly prominent where indigenous cultures had aided their establishment over the previous centuries.[7] The missionary reached the heart of French settlement in the region at Québec by traveling through the mixed landscapes of the upper Saint Lawrence that he found "agreeable in several places" and even "very pleasant."[8]

Sagard reveled in describing flora for his French readers.[9] In a later chapter of his 1632 *Grand Voyage* dedicated to the "fruits, plants, trees & riches of the country," his familiarity with the environments of this New France became clear. He confidently named and described cedar, oak, wild cherries, onions, raspberries, grapes, and lilies.[10] He was uniquely attentive to indigenous names and knowledge of these plants, but even in these moments he highlighted an essential similarity that transcended cultural differences in use and identification. If Sagard named the carnivorous pitcher plant (*Sarracenia purpurea*) "*Angyahouiche Orichya,* which is to say, turtle's stocking," for example, he included indigenous names more often as an opportunity to reassure his readers that these were simply new names for old plants.[11] He described, for example, "Cedars, named *Asquata,*" and "Roses, that they call *Eindauhatayon.*"[12] Sagard and the other French merchants, missionaries, and explorers who traveled to northeastern North America in the early seventeenth century were confident that they had discovered a not-so-New World, recognizable with existing names and amenable to French uses.

Sagard showed little disorientation, nor did he give any hint that the ecological novelty of North America might pose a significant challenge to Aristotelian intellectual traditions.[13] Instead, even if the geographical distance that the Récollet traveled was remarkable, it seems obvious that nowhere else could Europeans have traveled so far to find environments so familiar.[14] In northeastern North America, common ecological and evolutionary histories meant that the environments to which the French had come shared a great deal with those they had left.[15] American maples looked much like those of France (and the same was as true of plants as large and as common as birch as it was for more locally available strawberries, raspberries, and grapes), but they nonetheless differed substantially. We would now recognize them as different species of related botanical genera and families, but early colonists mapped these differences onto a distinction between the *sauvage* (wild) and cultivated. Close encounter and empirical engagement with these new places and new plants facilitated French efforts to plant an empire in seventeenth-century North America. The nature of New France was uncanny—simultaneously familiar and foreign—and naming the *sauvage* inspired confidence that this tension could be overcome and managed (Figure 3).

Figure 3. Louis Nicolas, "Sauvages plants," *Codex Canadensis*, ca. 1674–80.
Gilcrease Museum. Acc. 4726.7. Courtesy of the Gilcrease Museum, Tulsa,
Oklahoma.

Sagard's encounter suggests the need to weigh the effect of the cultural or mental "baggage" that explorers, colonists, and missionaries brought with them to the Americas in environments that shared common natural histories with Europe and that, in effect, were far more familiar than foreign for early modern explorers.[16] Sagard's lack of confusion can seem surprising if we expect European explorers utterly unmoored by the newness of the Americas. J. H. Elliott's argument that the impact of the New World on the Old was "disappointingly muted and slow in materializing" has encouraged a generation of scholars to debate the cause of a "blunted impact" that seems to echo in Sagard's confidence.[17] As the inheritors of decades of scholarly investigation of narratives of travel and exploration that in extremis claim that "we can be certain only that European representations of the New World tell us something about the European practice of representation," we are justly skeptical of authors such as Sagard, Champlain, and other settlers and missionaries in northeastern North America who demonstrate a considerable familiarity with what we know were new places.[18] At its most innocent, the failure of authors such as Sagard to recognize the novelty of American environments can seem a confusion of kinds and an inability to comprehend how the distance that they had traveled had translated into real biological and ecological differences.[19] Yet Sagard seemed reassured by what he found, and he saw in New France environments that were more French than new.[20]

Appreciating the colonialist intent of explorers such as Sagard who arrived with the explicit purpose of subverting indigenous cultures demands that we recognize that the power of Sagard's diagnosis of fruit that was "small for lack of being cultivated and grafted" or grapes that produced poor wine because local indigenous communities "did not cultivate them" worked through immersion in the natural worlds of early America rather than obfuscation of these material realities.[21] If, as Michel de Certeau has described colonial accounts, this was "writing that conquers," it was not because it "will use the New World as if it is a blank, 'savage' page on which Western desire will be written."[22] While a knowing confusion obscured indigenous title and promised easy European settlement in other parts of the Americas, in New France it was Sagard's efforts to foreground familiarity that instead justified European colonialism through close and precise study of American environments.[23] Empirical and accurate observation of plant morphology and ecology became essential acts in the establishment of a French colonial political ecology in northeastern North America that diagnosed the region as being in need of colonialism that its architects represented as a form of cultivation. Sagard's gardens, like his narrative more broadly, in this way reveal how French visions of American colonialism drew upon both material histories acting across millions of years and more recent cultural revolutions that enabled explorers and missionaries to appreciate them.

*   *   *

In his 1603 *Des Sauvages*, Samuel de Champlain recounted his first sighting of North America as an encounter not with land but instead with a "bank of ice that was more than eight leagues in length, with an infinity of smaller other ones." It was spotted April 29 when, their pilot estimated, they remained "one hundred or two hundred leagues from the land of Canada."[24] Champlain's voyage with a Spanish vessel to the West Indies only a few years earlier must have seemed a distant dream and the frigid lands he was discovering in this New France the near opposite of the terrestrial paradises to be found among the islands of the Caribbean.[25] Almost seventy years earlier in 1534, Jacques Cartier had a similar first impression of the northern reaches of North America, noting the "large number of blocks of ice along the coast" of Newfoundland. Before discovering the Saint Lawrence and more fertile lands to the west he found himself "inclined to believe that this is the land that God gave to Cain" near what is now Labrador.[26] Yet within a few short years these northern shores would be home to several tentative efforts at French colonization of the Americas. Early accounts of the region therefore attempted both to describe the frequently harsh and inhospitable settings and to hint at the possibility that the transplantation of French ecological regimes could ameliorate these extreme colds.

The Europeans who traveled to northeastern North America in the early seventeenth century found environments that resembled those they had left more than any others in the early modern world.[27] Many species of North American flora would have been almost immediately recognizable to European travelers who were familiar with European congeners, or related plants that shared a common genus. Birch, oak, maple, and other common features of temperate and boreal environments on both sides of the Atlantic were visible traces of shared geological and ecological histories produced by tectonic plate movements and climate change. This was equally true of smaller shrubs, fruits, and grasses.[28]

This was in fact a not-so-New World. French colonists recognized the flora of North America because it was made up of the same types of plants as the environments that they had left in France. Champlain was able to catalogue "oak, aspen, poplar, hops, ash, maple, beech, cedar, [and] very few pines and firs" so readily because of the existence of real botanical continuities between his Old World and the New and, more specifically, between northeastern North America and Europe.[29] These trees, along with the walnut and chestnut that Sagard described, and many other noticeable features of the contact-era environment were the descendants of once globally distributed plant communities that, across millions of years, circulated and spread throughout the northern hemisphere of North America, Europe, and Asia.[30] Scientific attention to the presence of related species of plants across massive distances and natural boundaries such as mountains and oceans—what biogeographers call disjunct distributions—has thrived for centuries.[31] More recently, with increased knowledge of geological processes and the science of plate tectonics, biogeographers have been able to write the history of the Earth's plant communities with much greater clarity. If considerable uncertainty remains around the specific timing and sequence of geological and evolutionary events, research into plant fossils, geological history, and molecular analysis has established a general paradigm.

In the past several decades, biogeographers have reconstructed the "widespread distribution of temperate forest elements in the northern hemisphere" during the Tertiary, or between 65 and 2 million years ago.[32] The forests that Champlain and Sagard left behind in Europe were the remnants of a much larger floral community that already included familiar trees such as maples, walnut, hornbeam, and sycamore, along with the ancestors of current species of *Prunus* such as cherries and plums.[33] They were more properly part of an "Arctotertiary geoflora" or "boreotropical flora" that can best be described as a "warm-temperate evergreen broadleaf and mixed forest" that had spread across temperate regions of the world that were much farther north than they are now.[34] These connections were more recent than the early supercontinents such as Pangaea or Gondwana and date instead to a series of connections between Asia, North America, and Europe that allowed for considerable communication of plants and animals between these continents.[35] To the west, a land bridge crossed what is now the Bering Strait and, to the east, volcanic activity created a land bridge crossing what is now Greenland, bridging North America and Europe.[36] The unique floras of local regions in the northern hemisphere were the product of repeated ruptures and re-creations of these bridges, as well as the accident of geographical features such as mountains, rivers, and inland seas.[37]

In the more recent past, global climatic cooling effectively broke these connections, driving plants and animals fast enough to adapt to the south and to isolated ice-free high points created as the ice sheets moved south and glaciers surrounded them.[38] In Europe, many tropical species were pushed to the south and died out as the Mediterranean blocked their escape.[39] Isolated by glaciers and oceans, North American, Asian, and European plant and animal populations began to diverge.[40] The floras of Europe and North America would have been most similar in the early Tertiary, before divergence began between 10 and 40 million years ago.[41] The specific makeup of individual ecosystems, shaped as much by geography as they were their specific organismal composition, drove evolutionary patterns and the emergence of new species of genera that remain common to much of the temperate northern hemisphere.[42] Many of the species unique to American and European ecologies therefore share common ancestors and visible similarities and, while different species, would have been familiar as types of plants that many of the French who crossed the Atlantic already knew.

However recognizable many parts of the local flora were, the cold was an unavoidable fact in the nascent colony, whether it was observed at Tadoussac, Port Royal, or Québec.[43] The lands that Champlain saw upon entering the Saint Lawrence River were "sterile" and lacking in any obvious "conveniences."[44] They were, he wrote, subject to "impetuous winds" and the "coldness that they brought with them."[45] It was a cold that once on land he found simply "excessive."[46] Archaeological and textual records from around the wider Atlantic world support these observations, suggesting that the first decade of the seventeenth century was one of the worst in a longer "little ice age."[47] Even if, as Champlain himself noted, the harshness of the winter varied annually, there is little doubt that these were years of immense climatological stress across the Americas and the Atlantic world.[48] Where a first French colony at Île Sainte-Croix in 1604 succumbed to the cold and scurvy, it was soon followed by the English colony of Sagadahoc; founded in 1607, Sagadahoc was itself a victim of a particularly harsh winter that was felt on both sides of the Atlantic.[49] Further to the south that same year, would-be settlers at Jamestown found themselves in the middle of the worst drought to hit the region in 770 years.[50]

It need not be surprising then that other early authors such as Marc Lescarbot and the Jesuit Pierre Biard expressed a deep ambivalence about the North American climate. Both authors traveled to and described Acadia within the first decade of its founding, and both lavished attention on the experience of Canadian winters. When, in 1604, Champlain arrived at Île Sainte-Croix,

"winter came upon us sooner than we expected."[51] It was a defining experience for Champlain as he sought to understand this region, and he wrote that
"it was difficult to know this country without having wintered here."[52] Within
a few short months and the abandonment of their first colony, the attention
of French colonial activity turned toward what would soon become Port Royal.
Here again winter proved a bitter experience. The Sieur de Monts had originally
hoped to move the struggling colony further south to "escape the coldness"
but was uninspired by the coast of present-day Maine and instead turned his
focus toward the Bay of Fundy.[53] Port Royal, it was hoped, would be "softer
and more temperate."[54] Champlain expected that this would put the colony
"at the shelter from the northwest," as well as provide a suitable site for the
agricultural colony imagined as gardens were laid out, fields were sown, and
buildings erected.[55] Although the winter of 1605–6 proved less "bitter" than
that of the year before, it nonetheless provided the background for another
vicious attack of the "mal de la terre," or scurvy.[56] It was a disease that attacked
like clockwork as groups of colonists settled New France, adding considerable mortality to the other inconveniences of an already difficult season. Even
if authors such as Marc Lescarbot presented Acadia as a potential "terrestrial
paradise," the harsh winters featured prominently in their accounts.[57]

There is little to suggest the sort of shock that intellectual historians suggest was common for travelers to the equatorial Americas, yet early settlers
nonetheless seemed surprised at colonial environments that challenged contemporary theories that privileged latitude as a primary determinant of weather
and as a key term in their geographical imaginary.[58] "There ought to be,"
claimed the Jesuit Biard, "the same sort of Climate in every respect as that of
our France."[59] This was, or could be at least, "a twin land with ours, subject
to the same influences, lying in the same latitude."[60] At Québec almost a decade later, the Récollet Sagard was still surprised that even if the settlement
was "by the 46th degree and a half [and] almost two degrees farther south
than Paris, . . . nonetheless the winter is longer and the country colder."[61]
Divided into five bands that included frigid, temperate, and torrid zones, the
model of the world inherited by explorers of the late fifteenth and early sixteenth centuries included regions that would necessarily be uninhabitable.[62]
Into the late seventeenth century, authors such as Nicolas Denys insisted on
the natural parallels between New France and Old that were signified by their
common latitudes. "All of this extent of New France contains only 5 degrees,"
he explained, situating the colony between Bayonne to the south and Calais
to the north "to make it understood that New France can produce all that

the old one can and as well, but it will require a great number of people's labor to bring it into productivity."[63] The temperate regions (defined most commonly and loosely as supporting life) that lay between arctic and torrid wastes could reliably be predicted to sustain the sort of life left by European colonists. The latitude of settlements in New France predicted climates and environments that would be similar to those across the Atlantic.[64]

Authors such as Biard and Lescarbot were participating in a broader re-conceptualization of cold environments taking place on both sides of the Atlantic.[65] Throughout Europe, scholars deduced that a rational God would not have created uninhabitable regions of the world. At the same moment, English expeditions to the north and Spanish and Portuguese explorations of the global south revealed sizable human populations that defied classical and medieval assumptions about climate, population, and culture.[66] The sixteenth century had seen massive human and capital investment in the exploitation of northern fisheries. Explorers followed suit and sought out minerals, potential settlement sites, and an elusive northwest passage throughout the century.[67] Encounters with people who made their homes in these bitter climates encouraged the tempering of long-standing beliefs in the inferiority of both northern places and peoples. The fact that the regions of the world that cosmographers identified as most amenable to settlement had already been claimed by Iberian powers pushed even skeptics toward support of colonization in the north.[68]

Even if they were at odds with Biard and other Jesuits over the direction of colonization in the region that we now know as Nova Scotia, New Brunswick, and northern Maine, secular authors such as the lawyer and colonist Marc Lescarbot wrote with the same faith in the possibilities for agriculture and colonial development in the region.[69] Each argued the limitations of sweeping assumptions about global climates and devoted considerable effort to understanding the specificity of the relationship between geography and weather. Biard suggested that "this country being, as we have said, parallel to our France, that is, in the same climate and latitude, by a principle of Astrology it ought to have the same physical forces, deviations and temperatures."[70] He called for close empirical study to determine the nature of local weather patterns when he wrote that "nevertheless, whatever the Astrologers may say, it must be confessed that that country (generally speaking, and as it is at present) is colder than our France, and that they differ greatly from each other in regard to weather and seasons. The causes thereof not being in the sky, we must seek them upon the earth."[71] Biard confidently looked to a future where

New France was recognizably French in climate, environment, and culture, and he understood the urgent study of colonial environments to be an inherently political project.

Confidence in New France's agricultural future remained as French attention focused westward after the founding of Québec in 1608, Trois-Rivières in 1634, and Montréal in 1642.[72] Accounts of harsh winters in a "locus horribilis" could seem designed to terrify readers but also affirmed the ultimate ability (and duty) of missionaries and colonists to conquer the season for their faith and their king.[73] French study of the area focused on temperate tendencies that could be accentuated and developed with colonial intervention. In the short term, explained seigneur and colonial promoter Pierre Boucher, the cold "does not impede anybody from doing what needs to be done; we wear a little more clothing than ordinary; we cover our hands with a type of muffle, called in this country mittens; we have great fires in our houses, because wood costs nothing here to chop and carry to the fire."[74] Adapting to these winters became a key facet of colonial experience and a principal ingredient in the articulation of colonial culture.[75] If the climate of North America presented these and other colonists with a puzzle, then it was one that these authors imagined would be easily solved for hardworking colonists who were trained to see such differences as superficial and transient.[76]

* * *

Images of American environments that envisioned discrete climatic bands gradually gave way to understandings of a more complex relationship between weather and geography. When Sagard, Champlain, and other early settlers of New France traveled to the Saint Lawrence Valley, they were drawn toward the landscapes of the Saint Lawrence platform: lowlands between the Appalachian Mountains to the south and the Laurentians to the north. This geological province largely begins at Québec, extending south along the Saint Lawrence and into the eastern Great Lakes region. It was a transition that was easily noticeable to early explorers. From "lofty mountains" at the mouth of the Saguenay, Jacques Cartier described finding himself among land "covered with fine high trees and with many vines," with "as good soil as it is possible to find."[77] This was, he suggested, truly the beginning of the area that he called Canada. The transition was as jarring for Champlain, who likewise traveled from a "very unpleasant land, as much on one coast as the other," at Saguenay to the region of Québec that was "beautiful and pleasant, and supports all sorts of grains to maturity."[78] These explorers noticed what geologists have since confirmed: the Saint Lawrence lowlands, once home to massive lakes and seas as glaciers receded, melted, and deposited rich alluvial clays that would support indigenous and colonial agriculture, was a temperate oasis in the midst of an otherwise forbidding landscape.[79]

These authors were describing an ecotone, or a transition between one climatic zone and the next.[80] The Saint Lawrence Valley was the northernmost reach of temperate North America and remained a mixed and ecologically complex region.[81] The forests that the French encountered were the product of geographical and climatic push and pull that exerted their force over millions of years. Periods of global cooling had pushed boreal elements such as pine and spruce south and temperate elements such as maple and chestnut had returned north with broader periods of warming that inevitably returned.[82] The colonists and administrators who settled New France therefore established their colony at the boundary between these two ecological regions. These are frequently particularly rich and diverse environments, home to more species than either ecological region individually. In practical terms, the ecotone offered the advantages of access to the resources of multiple ecosystems.

The mosaic-like quality of environments that blended boreal and temperate flora attracted the attention of colonial authors.[83] Beyond offering more resources, however, these rich and diverse environments also offered evidence for the possible future of New France. These were landscapes that were familiar enough to beckon colonists and that were yet different enough to demand their intervention. Rather than simply effacing the differences between European and American environments, the accounts of Champlain, Sagard, Lescarbot, and others of this first generation drew attention to them and offered the region's environmental history as evidence of a need for French intervention.

The natural richness and diversity of these spaces had been cultivated by millennia of indigenous occupation. As the French returned to the Saint Lawrence in the seventeenth century, a great deal had changed since it had been visited by Jacques Cartier. Where Cartier had visited thriving Iroquoian cultures that practiced agriculture and established large sedentary villages at sites such as Hochelaga (Montréal) and Stadacona (Québec), Champlain now found Algonquian peoples who traded for rather than cultivating corn, beans, squash, and tobacco. Each occupied and exploited ecological niches that supported their own lives and their commerce with each other. When Champlain visited what is now Montréal, he described the evidence of recent Iroquoian agriculture. He wrote:

> For higher up than this place (which we named Place Royale) at a
> league's distance from Mount Royal, there are many small rocks
> and very dangerous shoals. And near this Place Royale there is a
> small river, which leads some distance into the interior, alongside
> which are more than sixty arpents of land, which have been cleared
> and now like meadows, where one might sow grain and do garden-
> ing. Formerly the *sauvages* cultivated these lands.[84]

Champlain neglected to describe any features of Iroquoian agriculture that
might have persisted, but there is reason to believe that at least some of the
plants might have continued to grow into the seventeenth century. In the twentieth century, anthropologists reported finding wild tobacco at sites of previous indigenous occupation, and there is no reason to suspect that this would
not have been true at Montréal as well.[85] At the very least, the landscape that
Iroquoians had cultivated and that favored the population of local animals
such as deer and plants such as wild fruits continued to flourish in the "edge
effect" created by previous clearings of farm land and firing of local forests.[86]

Authors soon explained to their Atlantic audiences that the transitional
quality of this region was more relevant to understanding New France than
zonal bands that drew their significance from latitudinal transitions. The
Récollet Denis Jamet explained that an east-west axis changed abruptly at
Québec, writing,

> For the temperature of the air we find it similar to that of France,
> except for the heat which is more ardent to us. As for the winter I
> can only say what others have stayed here several years have said.
> The snows are larger than those of France and last ordinarily four
> months. The freezing is more violent (than in France) [and] the
> great river freezes up to the ocean. Something good is that we do
> not feel the cold winds like in France. From Gaspé up to Québec,
> which is almost two hundred leagues are only high and terrifying
> mountains fertile only in rocks and pines. But after Québec the
> lands are beautiful and have the possibility of being good if they
> were cultivated.[87]

The explorations of Champlain and others who followed the Saint Lawrence
traced the transition between boreal and temperate regions of the continent
and between the Canadian Shield, the Laurentian lowlands, and the Appala-

chian Mountains to the south.[88] In the "Carte géographique de la Nouvelle Franse" included in Champlain's 1613 *Voyages*, this ecological shift was marked clearly (Figure 4). The printed page itself clearly divided the terrestrial regions of the northeast along a north-south axis at the mouth of the Saint Lawrence. To the right, the landscape was mountainous with isolated groups of one to three trees. To the left, with the exception of specific landmarks such as "montreal," the landscape leveled out and was populated by large clusters of identical trees.[89] Farther west, explorers encountered a continental climate that was ameliorated considerably by the presence of the Great Lakes and regional variations that indigenous cultures had used to develop complex agricultural communities in the preceding millennia.[90] These authors were also aware that moving southwest along the Saint Lawrence accelerated these tendencies. Champlain described Haudenosaunee lands, for example, as "temperate, without much winter, or very little."[91] New France might therefore be inhospitable, but it was not at all uninhabitable to those who understood it properly.

*   *   *

Even as they would have vicariously experienced the severity of these early American winters, then, readers of the first French texts that described American environments would also learn about the climatic and ecological variability of these places. Champlain, for example, was quick to note that the winter of 1605–6 was not nearly as harsh as the one the year before.[92] Where colonial authors lacked adequate experience to judge the typicality of any given extreme, they could turn to indigenous peoples for a broader temporal range. In 1636, for example, Paul Le Jeune related that

> there was a great Northeaster accompanied by a rainfall which lasted a long time, and by a cold severe enough to freeze this water as soon as it touched anything; so that when this rain fell upon the trees, from the summit to the roots it was converted into ice-crystals, which encased both the trunk and the branches, causing for a long time all our great forests to seem but a forest of crystal,—for, indeed, the ice which everywhere completely covered them was thicker than a coin. In a word, all the bushes and everything above the snow were surrounded on all sides and encased in ice. The *Sauvages* told me that this did not happen often.[93]

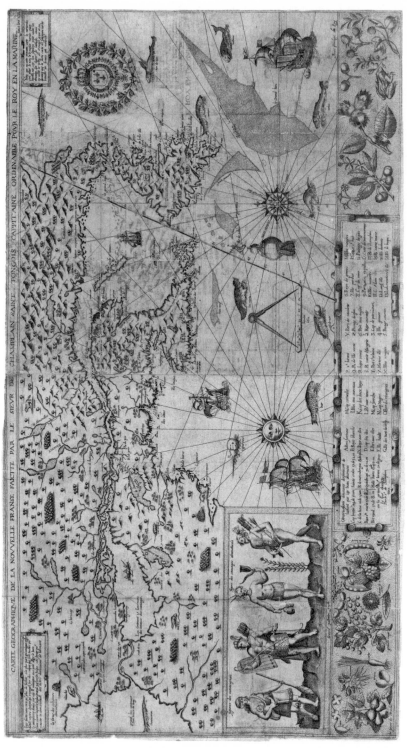

Figure 4: Samuel de Champlain, "Carte géographique de la Nouvelle Franse," *Les voyages du Sieur de Champlain*, 1613. Courtesy of the John Carter Brown Library at Brown University, Providence, Rhode Island.

Environmental and climatic outliers might inflict suffering or even kill, but they could safely be marginalized by presenting more temperate norms to metropolitan readers. This was an era of extreme climates on both sides of the Atlantic, and audiences could be relied upon to understand the difference between seasonally and locally variable weather and climatic patterns that emerged over a longer *durée*.[94]

These authors therefore paid exquisite attention to local ecological and climatological variation across New France. Pierre Biard explained how to properly read the clues presented by landscapes when he advised readers to note locations suitable for colonial development by the soil's "black color, by the high trees, strong and straight, which it nourishes, by the plants and grasses, often as high as a man, and similar things."[95] As Champlain returned to the Saint Lawrence Valley in 1608 with an eye toward establishing Québec, he noted a variety of landscapes and environments. Each setting was discursively composed of recognizable features that, whether present, absent, or noticed in place-specific combinations with other ecological markers, offered insight into what we now might consider the local ecology. Near Tadoussac, for example, Champlain saw only "mountains and rocky promontories . . . uninhabited by animals or birds." Even areas that at first seemed "the most pleasant" revealed only small, seasonally present birds.[96] Places such as "the island of hares" or "the river of salmon" memorialized ecological encounters and likewise made significant species or environmental features stand-ins for the environment as a whole.[97] Where there was no indigenous occupation that could testify to agricultural promise, trees became a particularly prominent feature of such accounts as they attested to soil qualities and other aspects otherwise hidden from view. The land between Tadoussac and Île d'Orléans, for example, showed only "some pines, fir and birch," evidence that this was truly "very bad."[98] As he passed the Gaspé Bay, Gabriel Sagard wrote that "all of this country is mountainous and high, almost everywhere sterile and unwelcoming, showing nothing but several pine, birch and other little trees."[99] As latitude could be a good predictor of what would grow, so too were the things that inhabited a place clues about its nature and the possibilities it offered to would-be cultivators.[100]

Explorers and early settlers highlighted the presence of familiar plants that were legible as symbols of the habitability—or potential habitability—of Laurentian environments. Selective inclusions of novel plants in their texts are therefore worth noting, but narrative accounts of settlement, exploration, and evangelization were not primarily interested in cataloging new botanical species.[101] Their authors instead populated environments with recognizable features and familiar names even if, armed with modern scientific taxonomies, we would name many of the plants that they encountered new species. Instances where the novelty of American flora was immediately recognized are thus rare in colonial texts. Champlain, for example, brought together a number of discursive and rhetorical strategies that limited the sense of cognitive disorientation that his French audience might have felt in the course of reading his numerous accounts of foreign cultural and natural environments. His description of Wendat territory in 1615 offered numerous representative examples of these features of early colonial texts.

> All of this country where I was contains some 20 or 30 leagues and is very beautiful, under a maximum of forty four degrees and a half of latitude, very well cleared lands, where they sow a great quantity of Turkish wheat [bleds d'inde], which grow beautifully there, as well as pumpkins, [and] sunflower, of which they make oil from the seed. . . . There are many vines and prunes which are very good, raspberry, strawberries, little wild apples, nuts, and a manner of fruit which is of the form and color of little lemons and which has none of the taste, but the inside is very good, and almost similar to that of figs. . . . It is a plant that carries them, which has a height of two and a half feet, each plant has only three or four leaves. . . . There are a quantity in several places, and the fruit is good and has a good flavor; oak, elm and beech, there are many spruce in this country. . . . There are also a number of little cherries and wild cherries [merise] and the same species of trees that we have in our forest of France are in this country. In truth the land seems a little sandy, but that does not mean it is not good for this type of wheat.[102]

In this fashion, the ecosystems of North America were presented as assemblages of recognizable plants. Embedded in descriptions of otherwise unfamiliar peoples and landscapes, the presence of recognizable French plants such as oak and grapevine anchored readers and travelers alike, offering promises of an essential similarity behind cultural and ecological difference more apparent than real.

Plants and places that French explorers and settlers were seeing for the first time in northeastern North America therefore rarely presented a disorienting challenge to Old World epistemologies or taxonomies. Colonial authors brought with them a transported taxonomy, a grid through which they learned about and made sense of novel American places. The names that an author such as Champlain used—broad classifications such as grain and herb, as well as specific names such as beech, oak, and maple—were all familiar European taxa, adapted and expanded to become abstract, generic terms. They became an implicit argument for a botanical unity between New France and Old that transcended the physical distance of the Atlantic Ocean. This meant that colonists and missionaries, employing the same terms that they used to describe French and European plants, incorporated American plants into a truly transatlantic flora that allowed them to feel familiar in environments that they were seeing and describing for the very first time. It also permitted French authors in North America to describe environments to European audiences in ways that mitigated distances and flattened the ecological and cultural specificity of what became a much more familiar New World.

Instead of working to catalogue individual species, a reliance on generic categories gave missionaries, colonists, and explorers considerable flexibility that allowed them to map Old World types onto American environments with a minimum of intellectual effort. Gabriel Sagard wrote in 1632 that "there are some pears, or that are called pears, certain small fruits a bit larger than peas, of a blackish color and soft, very good to eat."[103] Similarly, when the Jesuit Louis Nicolas wrote at the end of the seventeenth century that "the New World strawberry differs from ours only in that it is smaller, less fragrant and much more common," he was fundamentally relying on a common understanding of strawberry-ness that persisted in spite of changes in shape, color, smell, and distribution.[104] The result was to effectively mitigate the specificity of the plants they sought to describe, reducing morphological and ecological characteristics unique to North American populations to accidental traits that left essential characters unchanged. Therefore, when Jacques Bruyas wrote from his mission to the Haudenosaunee that he had found "walnuts and chestnuts, which I find in no wise different in taste from our own," the self-evidence of his identification was clear and the phrasing even redundant.[105]

These accounts imagined colonial landscapes populated with familiar plants such as pines, spruce, cherries, and raspberries, making new species of plants that differed morphologically and ecologically from those found in Europe portable and comprehensible outside of local sites of observation and experience in North America. Authors relied upon a body of knowledge shared with their European audience. What this meant was that as these authors noted the presence of oak, birch, or plum trees on the banks of rivers on which they traveled, in the environs of the missions at which they worked, and in the communities that they founded, they were drawing on a set of categories that would have been obvious to their intended readers. These were obvious when Lescarbot observed "cedar, fir, laurels, musk roses, currants, purslane, raspberry, ferns, lysimachia, a type of scammony, calamus odoratus, angelica, & other simples" at Port au Mouton.[106] They were equally obvious when Champlain saw forests "filled with woods, such as fir and birch."[107] The seemingly infinite references to trees such as pine and spruce or to French fruits such as cherries and plums testified to a confidence in the plasticity of common French botanical names that became, in effect, templates for the description and experience of American plants that were, as we understand them now, most often new species.

For both their present-day and contemporary readers, these references function by their obviousness. Named and listed as part of a floral catalogue of their new environments, these generic references can provide a crucial insight into both how French authors perceived North American plants and how they communicated their findings within the French Atlantic world and throughout Europe. This is in large part due to the overwhelming predominance of what scholars of folk taxonomies call a *folk generic* or *generic specieme*.[108] This, as ethnobotanist Brent Berlin writes, is the "category readily recognizable at *first glance*, as a single gestalt or configuration," and one that requires neither the use of specialized tools (i.e., microscope) nor considerable effort at differentiation.[109] Authors such as Champlain would have come with a set of generic floral templates based on their experience of the environments they had left. Having come from regions where there may have been just one example of a particular genus, all future species encountered were understood to be subtypes of this model. When Champlain commented on the birch he found in Wendake, for example, he identified the novel *Fagus grandifolia* with the *Fagus sylvatica* he would have known in France.[110] Differences in the shape of leaves and ecologies were effaced or minimized, and a piece of a new environment was cognitively domesticated.

With the use of these familiar names, the morphological details of a plant were most often simply implied. When he turned his attention to plants that he felt required more detail, Louis Nicolas situated them within particular cultural contexts and provided additional linguistic, medical, or economic information. When he described barley, for instance, he wrote that it was originally introduced from France and was used to make beer.[111] To describe a species of seaweed, he wrote only that small crustaceans survived the force of waves by growing filaments that kept them bound to the marine plant.[112] Thus even where a European botanical type was situated within a novel ecological or cultural context, its physical continuity with European plants was implied by relying on and reinforcing the salience of universally applicable types arrayed in lists compiled under a recognizable and familiar logic.

French familiarity was increased because of nearly a century of contact with American flora that had arrived as part of a broader Columbian Exchange.[113] American plants such as corn, pumpkins, sunflowers, and beans— widely sown in indigenous landscapes and a regular feature of travel accounts and natural history—were, by the beginning of the seventeenth century, relatively well-known in France. A recent analysis of the *Grandes heures* of Anne of Brittany, produced between 1503 and 1508, revealed the introduction, already by this early date, of a North American squash into cultivated landscapes in the Loire Valley. Botanical analysis of the "Quegourdes de Turquie" illustrated in the text suggests that it was an example of *Cucurbita pepo*, subspecies *texana* that was likely originally collected from the northern coasts of the Gulf of Mexico.[114] This is a variant of the same species that was represented in frescoes at the Villa Farnesina in Rome between 1515 and 1518, where the South American species *C. maxima* also appeared.[115] The eighteenth-century botanist Antoine de Jussieu also identified an American bean now known as *Phaseolus vulgaris* in the *Grandes Heures*.[116] How these plants arrived in the Loire Valley by 1508 and how they came to grow in the garden of Anne of Brittany, then queen of France, remains an open question, although the queen's strong personal ties to the papacy and Spanish crown may have inserted her into networks of botanical circulation that quickly diffused newly discovered American plants throughout Europe in the decades after first contact.[117]

There is little doubt that the squash in Anne's gardens were valued as a curiosity, but American flora became more economically significant and more widely grown as the century progressed. Charles Estienne, the author of the influential sixteenth-century *L'Agriculture et maison rustique*, wrote that "Turkish wheat [*blé de Turquie*], so called, or rather Indian wheat [*blé d'Inde*], . . . came originally from the west indies, then from Turkey and from there into France, not that it was cultivated for pleasure, or for the admiration of foreign things, of which the French give great weight."[118] Providing insights into the cultivation of the crop, he also offered advice on assimilating it into French lives. "It has a similar temperament to our wheat," he wrote, "always hotter, recognizable by the softness of the bread that is made with it."[119] Corn spread throughout Europe quickly in the wake of Spanish explorations of the Caribbean and

American mainland, although recent research into the genetics of European corn populations suggests that the spread of the crop into northern Europe awaited a second introduction from North America.[120] The diffusion of corn within France was slower than in contemporary Spain or Italy, but by the turn of the seventeenth century the crop was beginning to gain traction in rural regions such as Bresse.[121] Estienne likewise discussed the domestic cultivation of the pumpkins that Anne of Brittany had likely grown as a curiosity.[122] By the early seventeenth century, new editions of the *Maison rustique* also introduced tobacco as a valuable crop for landowners in France.[123] Champlain, Lescarbot, Biard, and Sagard had each left a France that was already home to many of the most widely cultivated American plants that they would find in New France.

*   *   *

The relatedness of French and northeastern North American environments was therefore frequently as obvious to seventeenth-century authors as it is to contemporary natural scientists. Accounting for this uncanny relatedness—a simultaneous recognition of difference and affinity—was *the* intellectual challenge for explorers, settlers, and missionaries who traveled and settled in the Saint Lawrence Valley and Acadia. It was neither possible nor preferable to completely ignore the differences that existed between European and American plants or the diversity that existed within North American plant populations. Adding additional information on cultural and religious significance or ecological and morphological distinctions, colonial authors modified generic botanical types to create what ethnobotanists refer to as *folk specifics* or *folk varietals*. This is obvious, for instance, when authors described white pine or red cedar. Assuming a shared body of characteristics (pine-ness or cedarness, in this case), authors could incorporate new flora with the greatest possible economy of description and communicate more effectively with their French audiences. The Jesuit Louis Nicolas wrote that there were three *façons* (later using the word "species" as well). The smallest type was not even given a name, and as for the difference between the red and white species, he wrote that they "differ only in the color of their bark."[124] While there were moments where these comparisons seem infelicitous, such as when Sagard described the Tupinambour (Sunchoke or Jerusalem Artichoke) as the "apple of Canada," they more often pointed to an acknowledgment of real botanical continuities that we now use evolutionary science to explain.[125]

Contact with indigenous cultures could provide colonists and explorers with new names for American plants and knowledge about their possible uses. Lescarbot related, for example, the vain search for the plant *Annedda* that had—a century earlier—cured Jacques Cartier's men of scurvy as an instance where this dependence meant death. "As to the tree *Annedda* to which Cartier has made mention," he wrote, "the *Sauvages* of these lands do not know it at all."[126] Cartier had described the miraculous plant only as "large and as tall as any I ever saw" but lavished attention on its powers and, it seems, brought specimens of the tree that were soon growing in the gardens at Fontainebleau. It "produced such a result," he wrote, "that had all the doctors of Louvain and Montpellier been there, with all the drugs of Alexandria, they could not have done so much in a year as did this tree in eight days."[127] The most obvious lesson from the incident for seventeenth-century colonists was to turn to indigenous peoples and ask for the curing plant. Champlain, when he discovered a Native near Tadoussac with the name "Aneda," seemed sure that "by this name was the one of his race who had found the herb *Aneda* known," even if "the *sauvages* do not know this herb at all."[128] Many of these truly novel plants, however, were presented to European audiences without names, such as those presumably edible roots that featured on Champlain's 1613 map.[129]

The use of indigenous names did not, however, universally imply a respect for indigenous knowledge or a lasting connection to specific indigenous communities. *Atoca* (cranberry), for example, was known by variants on this Wendat name after it was first described by Gabriel Sagard, but the name became generalized and lost its association to any specific community.[130] This missionary first transcribed the name as "toca" and wrote that with "neither pit nor seed, the Hurons [Wendat] eat it raw, and also put it in their little loaves," demonstrating the interwoven nature of botanical and cultural exchange.[131] The Jesuit Paul Le Jeune also recorded the fruit among the Haudenosaunee. He explained to his readers that "the young people went to gather it in the neighboring meadows, and, although it is neither palatable nor substantial, hunger made us find it excellent. It is almost of the color and size of a small cherry."[132] It was not just the French who appreciated the fruit. Le Jeune's confrère Louis Nicolas wrote that English colonists used the plant in place of verjuice, which was normally produced from unripe grapes.[133] Over time these references to the indigenous peoples who harvested this plant would decline, but the name, standardized as *Atoca*, remained the same.[134] The engineer Gédéon de Catalogne wrote that it was used to make preserves in 1712.[135] Antoine-Denis Raudot added that it was useful against dysentery, and the

Jesuit Pierre-François-Xavier de Charlevoix suggested its use for digestive ailments.[136] Orthography, descriptions of morphology, and the expected effects of the plants became fixed as names such as *Atoca* became part of a French Atlantic taxonomy. Linguistically, plants such as this retained their linkage to American soils, but, like the originally indigenous words "barbecue" and "canoe," their connection to any specific indigenous language was severed.[137]

*   *   *

Among multiple strategies for naming the simultaneous familiarity and foreignness of plants and places, describing the nature of New France as *sauvage* became a particularly powerful tool that both affirmed similarity and stigmatized difference. In many cases, morphological differences between European and American species of plants were considered red herrings, more apparent than real. When colonial authors described various American plants such as lemons, cherries, and oats as *sauvage*, they implicitly suggested that American flora was an imperfect or degraded version of that which existed in France. A common refrain that emerged as early as the writing of Champlain was that French agricultural techniques had brought a full expression of botanical essences in French plants that were only latent in their American kin. In studying grapevines, for instance, authors such as Champlain understood that the grapes that produced bitter and unremarkable wines from Louisiana to Acadia were a product of aboriginal neglect.[138] In his 1603 *Sauvages*, Champlain wrote that Québec contained "wild fruit trees, and vines: in my opinion, if they were cultivated they would be as good as ours."[139] In 1668, Jacques Bruyas wrote similarly about the vines near his mission among the Haudenosaunee at Saint François-Xavier (Kahnawake): "I believe that, if they were pruned two years in succession, the grapes would be as good as those of France."[140] The authors who described the *sauvage* plants of French North America focused on these subtle differences and diagnosed a lack of cultivation—of an unmet and unexplored potential in American plants. Where, throughout Louis Nicolas's *Histoire naturelle* and the *Jesuit Relations*, plants such as cherry trees or vines are identified as *sauvage*, they were speaking to this sense of an inferiority less innate than accidental.[141] Champlain, Bruyas, and Nicolas each claimed that the observable differences of American plants were mutable and credited them to the influence of the local North American environments and aboriginal ecological practice.

The term *sauvage*—as a noun—referred to the aboriginal communities of North America in the seventeenth century. The American *sauvage*, explains historian Olive Dickason, blended "the well-known Renaissance folkloric figure of the Wild Man; early Christian perceptions of monkeys, apes, and baboons; and the classical Greek and Roman tradition of the noble savage."[142] Aboriginal communities, as *sauvages*, were said to blur the line between civilization and savagery so that they lived in a perpetual state of wildness, more non-human than human in their customs and relationships with the natural world. If it did not carry many of the pejorative connotations of unrestrained violence that the English translation of "savage" does today, the characterization of aboriginal peoples as *sauvage* encouraged and justified the establishment of a French presence throughout North America as a project to reclaim and rehabilitate a degenerate people. It was a term therefore that could equally be applied to the human and non-human world and that suggested a deviation from the norms that defined the civilized French world.

It was a complex term that, already by the end of the sixteenth century, rejected a neat teleology or morality.[143] By the time of the colonization of Acadia and the Saint Lawrence, the term embodied some of the tension that would come to the fore as visions of the "noble savage" were articulated in eighteenth century. The author of a Jesuit journal, for example, wrote that "it is true that one lives in these countries in a great innocence."[144] Passages such as these seem to hint at the continuation of sixteenth-century discourses that evoked respect for the rustic and simple lives of indigenous peoples.[145] Famously, Michel de Montaigne wrote that "they are *sauvages*, just as we call *sauvages* the fruits, that nature, in itself and its ordinary progress, has produced: although, in truth, it is those that we have altered by our artifice, and turned away from the common order, that we should rather call *sauvages*. In the former are alive and vigorous the true and most useful and natural properties, which we have bastardized in the latter, and have accommodated them only to the pleasure of our corrupted tastes."[146] The use of the term *sauvage* therefore mapped closely onto debates about the distinction between the artificial and the natural and, more broadly still, about whether the natural world could be fundamentally improved upon.[147] Ultimately, however, one finds little of this sophistication in colonial texts that instead saw *sauvage* places and peoples as less than their French counterparts.

The ideology of the *sauvage*, as it was framed in New France, was ambivalent about any claims that the *sauvage* state of North America made it naturally superior to the culture that the French could introduce.[148] Instead, authors who named the aboriginal peoples of French North America *sauvage* encouraged treating them in a similar fashion to the wild grapevines that, one explorer wrote, were "lacking only a little culture."[149] If the implication of this language was a sense that aboriginal cultures were not irreconcilably different or inferior, it suggested that, under the right conditions, they too could be "cultivated" through their encounters with French missionaries and colonists.

Yet in practice colonial authors frequently recognized indigenous improvement of American environments. Sites with histories of indigenous occupation were often sought out for colonial settlement even where indigenous ecological knowledge was otherwise marginalized and dismissed. Studying the environment for hints of its habitability meant observing the ways in which indigenous cultures had lived with them and implicitly recognizing the merits of indigenous technologies and ecological practice. It is likely that early explorers were particularly attracted to land that had, in the preceding century, been home to indigenous communities.[150] One of the first farms established as the settlement at Québec grew, for example, has a documented history of aboriginal occupation that stretched back millennia and that had only declined with the broader disappearance of Iroquoian agriculture in the area during the sixteenth century.[151] Perhaps not surprisingly, in an era that regularly saw starvation threaten colonial settlements throughout the Americas, early explorers also focused a great deal of attention on the native foods of the region when they investigated Native cultures.[152] Observation of edible plants was intimately associated with French experience of indigenous peoples and their cultures. In the fields of the Iroquoian Wendat in what is now Ontario, for example, French efforts to ascertain the civility of the people meant studying the botanical company that they kept and carefully cataloguing the cultivation, preservation, and consumption of food crops.[153]

Indigenous ecological knowledge was both foregrounded and critiqued within accounts that drew upon the sylvan symbolism of the term *sauvage*. Missionary and colonial sources were clear that these *sauvage* people were unable to properly use *sauvage* plants to better themselves; they languished together. Lescarbot complained about the resistance of Mi'kmaq with whom he interacted to appreciate the evident superiority of French practice. "We showed them," he wrote, "in pressing grapes in a glass, that this was how we made the wine that we drank. We wanted to make them eat the grapes, but having them in their mouths they spit them out, and thought (as Ammianus Marcellinus recounts of our old Gauls) that it was poison, such are these people

ignorant of the best thing that God has given to man, after bread."[154] These
were critiques that therefore slipped readily between accusations of cultural
and moral inferiority and that rendered discussions of indigenous practices
evidence of the need for spiritual reform and civilization. Even if "the forest
serves" the indigenous peoples of New France, as one Jesuit explained, it was
because "they know better the ways of these vast and dreadful forests than do
the wild beasts, whose dwelling they are; the French did not lightly venture
to entangle themselves in these dense woods."[155]

Indigenous people in these accounts seemed both part of and subject to
the natural environments of New France. Champlain, for example, framed
his struggles to traverse "thick woods" while "loaded down with a pikeman's
corselet" and attacked by "hosts of mosquitoes, a strange sight, which were so
thick that they hardly allowed us to draw our breath," as a brutal episode saved
only by two aboriginal people who were simply "traversing the woods."[156] Elsewhere he described a waterfall that his guides crossed easily without getting
wet.[157] The Jesuit Paul Le Jeune was astonished that even as he experienced a
cold "so violent that we heard the trees split in the woods," he was visited by
indigenous peoples "sometimes half-naked, without complaining of the
cold."[158] Perhaps this was because, as Le Jeune later related, these Innu peoples
conceived of the seasons as non-human beings with whom they could interact; Pipounoukhe (winter) and Nipinoukhe (spring and summer) each shared
the world and could be heard "talking or rustling, especially at their coming."[159] Accounts such as these did not represent indigenous knowledge as the
product of skillful adaptations to harsh environments. Instead, colonial authors
marginalized complex technologies and skills as unlearned and unrefined
reactions of *sauvage* cultures.

As often as they seemed to remain above the material constraints of the
natural world that so challenged French exploration and settlement, both Iroquoian peoples to the west and Algonquian communities to the north were
represented as being mastered by, rather than masters of, their environments.[160]
Champlain, for example, compared the ordinary indigenous preparation of
dried fruit for winter to the practice of Lent in France; the power of absti-
nence during Lent is of course its voluntary nature but here in a land that
seemed capable of supporting great ecological diversity, such a fast was a
necessity.[161] Even as the seasonality of the continent's climate encouraged his
vision of an agricultural empire, the frequent migrations of indigenous communities such as the Odawa berry gatherers who Champlain encountered
in 1615 seemed dependent upon fruit that was "manna," or a gift from God,

rather than the product of their own labors.[162] Even the Wendat peoples with whom Champlain had frequent cause to winter—admirable agriculturalists though they were—were found lacking. "Their life is miserable when compared with ours," he wrote, "but happy among them because they have not tasted better."[163] Land was cleared "with great costs" and labored by women, and it produced dishes that "we would give to pigs to eat.[164]

In this way, *sauvage* became a term that as both noun and adjective described and explained the uncanny character of northeastern North American places and peoples. As an adjective, the *sauvage* laid claim to an essential biological equality between the flora and environments of New France and Europe. As a noun, it argued that indigenous peoples were unable to claim the requisite distance from their natural world required of civilized people. Together, they diagnosed place and people in tandem. Both were in a state of wildness, subject to the excesses and insufficiencies of the other.

*   *   *

When early authors—colonists, missionaries, and explorers such as Gabriel Sagard, Samuel de Champlain, Marc Lescarbot, and Paul Le Jeune—arrived in northeastern North America, they were drawn to a natural world that they called *sauvage*. They carefully recorded each encounter and experience with new places and new plants, arguing that, adequately understood, the nature of these regions would provide insight into how French colonialism could take root there. The political ecology of French colonialism in seventeenth-century North America translated the region's distinctive environmental history into evidence for the need for French intervention. In the same breath, colonial authors highlighted both affinities and differences between European and American environments. Floral and ecological similarities offered proof of an essential resemblance and unity, while real differences legitimized French efforts to marginalize indigenous ecological knowledge and ignore the sovereignty of the aboriginal communities who had long lived in what soon became New France.

The first narratives of exploration, evangelization, and settlement therefore moved consciously toward accounts of the region that were both human and natural histories. Gabriel Sagard, when he delighted in the gardens of his order at Québec, understood French intervention in northeastern environments as the fulfillment of a providential history that offered the promise that the *sauvage* nature he described could be perfected. Champlain, when he described early efforts to cultivate American grapes at his *habitation*, suggested that French colonialism would tap the region's unfulfilled potential. Close study and careful attention to the distribution of temperate flora provided proponents of French colonialism with a purpose for New France. As letters, narratives, specimens, and samples crossed the Atlantic in the first decades of the seventeenth century, understanding the environments of northeastern North America became central to France's colonial project.

CHAPTER 2

Communicating Cultivation

When Marc Lescarbot recounted his own experiences during the first years of Acadian settlement, he wrote from Paris with the benefit of hindsight and a lingering sense of loss. Both are obvious from his frequent and loving accounts of farming in northeastern North America. We know, for example, that the settlement on Île Sainte-Croix in what is now Maine lasted one harsh winter, but for Lescarbot what was worth remembering was "the nature of the land." It was, he promised, "very good and pleasantly abundant." He knew this from his own experience because their leader, Pierre du Gua, the Sieur de Monts, "had ordered several sections of land cultivated there" that soon revealed its promise—its nature—and demonstrated the power of French labor to plant agricultural lands that produced "marvelously."[1] The apothecary Louis Hébert had planned to work with local grapes at Port Royal, and, at what is now Canso, Nova Scotia, Lescarbot cultivated "his garden of wheat as beautiful as one knows in France."[2] The moral and legal legitimacy of French colonialism was established through efforts to turn over soils, to prune and graft indigenous and introduced flora, and to announce the success of French horticultural practices in a New World.

Lescarbot's account synthesized a century of French experience in North America but routinely highlighted the power of moments such as these where he or other colonists and explorers had worked closely with American environments. Cultivation was "likely the only innocent vocation," Lescarbot wrote, and failure to embrace it had doomed the French in Florida and the Iberian powers whose empires in the Americas remained instead extractive and exploitative.[3] Agricultural labor, we learn, was valued as much in classical tradition as it was commanded by God. Clearing land was dangerous but pleasing work that opened dangerous airs that had been trapped in the soil but also promised pleasures that made him confident that "he would never return to France."[4] It was through these efforts that Lescarbot learned the local climate, the proper seasons to plant, and the best means to realize the potential of local environments.[5] Cultivation was, in effect, the means through which Lescarbot came to understand the continent to which he had come and to appreciate the transformation that French colonialism could reap in American landscapes and indigenous lives.

Lescarbot's *Histoire de la Nouvelle-France* was an enormous book, but he remained consistent in his argument that the lands that French colonists had claimed in northeastern North America were uncultivated and *sauvage*. He acknowledged the differences between France and the country to which he had come. He found particular inspiration in Deuteronomy, where Moses had explained to his followers that "the country to which you go to possess is not like Egypt from which you are leaving, where you have sown your seeds and watered with the labor of your feet." Instead, "the country to which you go to possess is a land of mountains and plains and is watered as it pleases heaven."[6] Lescarbot therefore represented a landscape in natural simplicity, labored only by a beneficial God and awaiting a people who would come to take and improve upon the gifts that God had seen fit to bestow. He transmitted his experience of this place in a detailed narrative that, while routinely citing classical and contemporary authorities, foregrounded the knowledge acquired through the labor of his own hands and cemented as he reflected upon his close encounter with American environments.

As both place and people were diagnosed as *sauvage*, French plans for their salvation were inseparable parts of a worldview that, anachronistically, we might call ecological. Colonization, conceptualized as a form of cultivation, would draw out the potential of every facet of American environments. The scale and ambitions of French colonialism increased with its geographic footprint. In 1634, French settlement pushed west to Trois-Rivières, and in 1642 Montréal was established where over a century earlier Jacques Cartier had gazed upon an Iroquoian village. Colonists followed in the wake of missionaries who had traveled westward to establish themselves among Great Lakes communities within a decade of the founding of Québec. When, in the late seventeenth century, the explorer Henri Joutel described the western country that he had discovered as a participant in the explorations of René-Robert Cavelier, Sieur de La Salle, he wrote tellingly that "one finds there vine that is lacking only a little culture."[7] His call found echoes in other writings by missionaries and colonists and reveals the extent to which a French colonial political ecology that transmuted the visible differences between American and European flora into the difference between the *sauvage* and the cultivated shaped environmental encounters across seventeenth-century North America. Like Lescarbot, Joutel was participating in a broader theorization of empire itself as a form of cultivation, as a rehabilitative enterprise in early America that targeted both the human and the natural worlds.[8]

Making the case for cultivation required forms of communication that transported French audiences into the gardens and missions of northeastern North America. These were not claims that could be communicated through botanical specimens, nor could they be captured through the presentation of rare species of novel plants or animals to Parisian collectors. It was one thing to note the biological affinities between temperate ecosystems and flora that shared evolutionary histories, but the transition to identifying the environmental differences between New France and Old as remediable defects demanded rhetoric that foregrounded the firsthand experience of the authors who made the claim and narratives that could simultaneously describe the natural world and confidently diagnose it as deficient. While there was much about which Lescarbot and other colonial authors disagreed, they shared a common understanding that the act of cultivation would reveal the true—French—nature of the places and peoples of northeastern North America. It was an active process that demanded strength of character and clarity of purpose, but it was an enterprise that had been commanded by God and that promised real pleasure for those willing to take up the task. The cultivated spaces of French North America became opportunities to display a stewardship associated with the management of a landed estate.[9] This was an ideology that therefore valorized environmental practice as much as it offered a convenient metaphor to conceptualize the conversion of indigenous peoples; it authorized the claims of authors such as Lescarbot through a celebration of their labors and rooted claims of French sovereignty through the imposition of European horticultural regimes in American landscapes.[10]

The cultivated spaces of Acadia and the Saint Lawrence Valley in this way became privileged sites from which to lay claims to know and to own New France, but the success of these claims was dependent upon the means through which they were communicated to France. The rhetorical strategies of explorers such as Champlain, missionaries such as Sagard, and colonists such as Lescarbot did not rely upon an abstracting science but on immersive forms of writing that mediated the intellectual and geographical distance that separated New France from the Old. They presented North American environments as complex wholes best knowable through their own lived experiences. Just as the landscapes that they cultivated promiscuously blended elements of introduced and indigenous flora as witnesses to their own ability to channel the productive energies of colonial soils and environments, they were similarly promiscuous in their choice of forms and genres in which they wrote. Across the seventeenth century, the authors who communicated New World environments to French audiences did so in travel narratives, administrative documents, natural histories, and modes that blurred distinctions between personal accounts, colonialist propaganda, and protoscientific genres. In fact, the only consistent feature across these texts was a focus on emphasizing the essential familiarity of New World places and certainty in the promise that cultivation would produce a New France in northeastern North America.

*   *   *

As Samuel de Champlain expanded the French presence in Acadia and the Saint Lawrence Valley in the seventeenth century, he frequently created experimental sites to test the planting of French species and ecological practices. In both written texts and accompanying images, cultivated spaces functioned as visible beachheads for European ecological practice and species. As the *habitation* at Québec was being built in 1608, Champlain "had all the rest cleared so as to make gardens in which to sow seeds to see how they succeed."[11] He continued experimental plantings as he turned his attention west in the following years. In 1611, while waiting for the aboriginal guides who would lead him into the interior, he wrote that "I had two gardens made, one in the prairie and the other in the woods . . . and the second of June I sowed some seeds that grew in all perfection and in little time, that demonstrated the bounty of the land."[12] Even the crops of failed colonial endeavors such as those at Île Sainte-Croix offered evidence of the promise of the region—and the need for French colonization.[13] French travelers and colonists therefore drew a mandate from their experience cultivating northeastern environments.

Cultivation implied specific ecological practices as well as gesturing toward a larger organizing ideology. It meant, for example, clearing the land of woods and opening the soils to the warming sun through French labor. Gold and silver might be found, wrote Lescarbot, but "the first mine to have is bread and wine, and livestock."[14] Le Jeune explained this necessity was also an obligation when he wrote that "New France will someday be a terrestrial Paradise if our Lord continues to bestow upon it his blessings, both material and spiritual. But, meanwhile, its first inhabitants must do to it what Adam was commanded to do in that one which he lost by his own fault. God had placed him there to fertilize it by his own work and to preserve it by his vigilance, and not to stay there and do nothing."[15] Even if accomplishments were admittedly modest in the first decades of French colonization, with the colonization of Acadia suffering repeated setbacks during intermittent power struggles and residents of Québec having cleared only "18 or 20 acres at the most" by 1627, early advocates of both mission and colonization looked to early crops and assessments of American environments as evidence of a bright agricultural future.[16] Colonists and missionaries cleared land, sowed crops, and learned the seasons. In such a manner colonists came to understand what one Jesuit referred to in 1643 as "the spirit [*génie*] of this place."[17]

Gardens of necessity, those that fed colonists and established a visible claim to only newly settled lands, were also pleasure gardens.[18] As Lescarbot explained, "I can say without lying that I have never worked my body so hard, for the pleasure that I took to lay out and cultivate my gardens . . . to make *parterres*, to align the *allées*, to build the *cabinets*, to sow wheat, barley, oats, beans, peas, garden herbs and to water them, such was my desire to recognize [*reconoitre*] the land through my own experience."[19] For as much as the horrors of scurvy and frigid isolation shaped the experience of early colonists such as Lescarbot and Champlain, the simple act of setting roots—both figurative and real—in New France was a reassuring one. Cultivated spaces became sites in which to experience a harmony only possible as the result of a concerted effort by skilled husbandmen and an experience of desire and pleasure made possible by the fulfillment of a duty to cultivate place and people.

Lescarbot was particularly concerned with foregrounding this "little labor" that "God has blessed" in his accounts, but he was not unusual in his efforts to claim that agricultural labor provided an authority that was both moral and epistemological. As the Protestant polymath Bernard Palissy explained in the sixteenth century, the labor of cultivation was inspired by both God and Roman precedent. It was in the classical era that people had

> wisely set themselves to plant, sow and cultivate to aid nature, which is why the first inventors of something good, to aid nature, have been so esteemed by our predecessors, they were reputed to have been participants in the spirit of God. Ceres who advised us to sow and cultivate wheat was called a goddess; the good man Bacchus (not at all a drunkard as the painters have made him) was exulted

because he advised us to plant and cultivate the vine: . . . Bacchus had found *sauvage* grapes, Ceres had found *sauvage* wheat; but these were insufficient to feed them as well as when they were transplanted. From this we know that God wants us to work to aid nature.[20]

It was through labor that the productive essence of *sauvage* plants was revealed and the foundations for European civilization laid.

Authors were inspired by developments in the evolution of renaissance gardens that created spaces in which the natural and artificial were intentionally blurred to demonstrate the skillful labor of a benevolent patriarchal authority.[21] In a description worth quoting at length, Champlain hinted that rich landscapes that aimed to blend human and natural agencies were as important in North America:

As soon as the said Sieur de Monts had departed, some of the forty or forty-five who stayed behind began to make gardens. I also, in order not to remain idle, made one which I surrounded with ditches full of water wherein I placed some very fine trout; and through it flowed three brooks of very clear running water from which the greater part of our settlement was supplied. I constructed it near the seashore a little sluiceway, to draw off the water whenever I desired. This spot was completely surrounded by meadows, and there I arranged a summer-house with fine trees, in order that I might enjoy the fresh air. I constructed there like-wise a small reservoir to hold salt-water fish, which we took out as we required them. I also sowed there some seeds which throve well; and I took therein a particular pleasure, although beforehand it had entailed a great deal of labor. We often resorted there to pass the time, and it seemed as if the little birds thereabouts received pleasure from this; for they gathered in great numbers and warbled and chirped so pleasantly that I do not think I ever heard the like.[22]

Agricultural and horticultural authors in the sixteenth century had laid claims that these sorts of labors were particularly virtuous, translating an act that could be conceptualized as one of Adam's punishment into a site where God could be encountered and where His designs could be made legible, and where one could expect rewards of pleasure and beauty.[23] These were places, therefore,

of both pleasure and toil, and accounts of these landscapes demonstrated the
possibility of creating the "third nature"—a skillful display of the subtle manipulation of natural agencies—that renaissance and early modern gardens
championed.[24]

In this way, colonial narratives framed French colonialism less as imposition than as redemption and extended a project that had thus far targeted
the environments of rural France for improvement into the Atlantic world.[25]
Imagining effective husbandry as an encounter that cultivated farmer and
farmed alike, it encouraged both confidence in the ability of French colonists to weather the influence of New World environments and a belief that
civilization would allow for the expression of latent identities—a possible
Frenchness—in both place and peoples. It offered French colonists and colonial promoters an image of themselves and their project that differed substantially from the bloody conquests of New Spain and Peru and that allowed
them to imagine sovereignty over New World possessions and control over
American peoples as recompense in a consensual and mutually beneficial exchange of lands, goods, and cultures.[26] For the cultivator could not be a conquistador, and the manifestation of his (for this figure was invariably male)
will was imagined as operating far more subtly, offering a glimpse at an
understanding of the importance of the agency of both the cultivator and
the cultivated to the success of colonization.[27]

Narratives of cultivation in New France translated French practices of
mesnagement, or stewardship, flourishing in France across the Atlantic and into
American soils.[28] In France, the reclaiming and improvement of agricultural
and horticultural lands by French cultural and political elite was part of a
broader strategy of political centralization that produced a robust visual and
material culture of royal authority in early modern France that drew upon the
metaphor of a benevolent patriarch who enriched both his own holdings and
his subjects.[29] These produced frequently formal landscapes that were sites of
political practice where new visual and practical forms of power were articulated and experimented.[30] Texts on gardening and estate management had
begun to proliferate in France after the middle of the sixteenth century following military adventures in Italy, both introducing classical authorities on
the subject and developing a French aesthetic and landscape theory.[31] During
this period, gardeners and landscape designers hired from Italy produced gardens such as Fontainebleau and invented the French mannerist style.[32] The
French formal garden was born in the seventeenth century in gardens such as
the Tuileries and the Luxembourg palace through royal support and horti-

cultural innovation that emphasized the aesthetic beauty of purposeful design and utility.[33]

In France and England, authors were reengaging the georgic tradition of Virgil and other classical authors that valorized the active management of estates by their noble owners.[34] The horticultural theory that emerged in late sixteenth-century France celebrated the empirical roots of botanical knowledge acquired through the virtuous labor that expanded patriarchal authority in French rural landscapes. In the 1600 *Théâtre d'agriculture et mesnage des champs* of Olivier de Serres, a leading horticultural theorist patronized by Henri IV, we can see how the practice of *mesnagement* brought the disparate concerns of epistemology, agricultural improvement, and authority into a productive relationship. Serres proposed understanding agriculture as a science that observed and adapted to local environmental conditions.[35] It was a "science more useful than difficult, provided that it was understood by its principle, applied with reason, led by experience and practiced with diligence."[36] Serres had escaped the ravages of France's wars of religion on his estate and had there learned practices for effective estate management that he felt were a moral obligation ordained by God. "For as much as God wants us to content ourselves with the places that he has given us," he wrote, "it is reasonable to take them from his hand as they are and serve him the best as possible trying to remedy their defaults through artifice and diligence."[37] What he developed at his estate could be applied to the state, with the king conceptualized as a *bon ménager* of the kingdom. These writings were particularly resonant in a France still feeling the effects of economic and cultural disruption that had characterized much of the sixteenth century.[38] Particularly relevant to the ambitions and concerns of Henri IV, Serres drew upon classical texts to argue that the management of people and the management of place were twinned and inseparable.[39]

Serres's calls for a nation of nobles who understood their role as effective managers of their estates, influenced heavily by readings of classical authorities such as Virgil, mapped neatly onto a larger seventeenth-century debate about the relationship between France and its Roman past. Sixteenth-century French humanists had claimed the Gauls as their colonized ancestors to chart a new intellectual course distinct from an inherited Roman past.[40] French theorizations of empire were heavily inspired by a Roman model (as all European nations were, to some extent).[41] As French intellectuals debated their cultural debt to Rome, an agricultural discourse enabled them to refigure differences between Roman and French culture (otherwise often understood to

be deficiencies in the French) into an evolutionary model that figured cultivation as a principal civilizing act.[42] Even if colonialism could occupy an
ambiguous place within the thought of Serres and his adherents—the Duc de
Sully, for example, saw colonialism as a drain on French population and an
overextension that defeated the effective management of French territories—
in stepping into the role of their former Roman colonizers, colonial promoters in France argued that their nation would itself be bettered.[43] Colonialism
would work alongside the cultivation of French arts and sciences to produce
a nation that surpassed its classical counterparts.[44]

*   *   *

Lescarbot's *Histoire* described a particularly momentous presentation of American flora to the French court that had taken place in 1607. This had been a
crucial moment only three years after the first colony in Acadia had been
founded and only two years after Port Royal had been established in 1605.
The colonists—Lescarbot among them—had just recently learned that their
monopoly on the fur trade that had been granted to their benefactor and leader,
the Sieur de Monts, had been revoked in the face of opposition from Atlantic
merchants in France.[45] In August of that year, the colonists at Port Royal had
been forced to return to a king and a court that seemed uncertain about how
or even whether to support colonization in North America. It was Jean de
Poutrincourt, a close friend of de Monts and lieutenant governor of his colony
in Acadia, who "presented to the King the fruits of the land from which they
came" when they arrived in Paris. Alongside geese and other fauna from North
America, it was "the grain, wheat, rye, barley, & oats, as the most precious
things that one could bring back from any country," that he delivered to king
and court.[46]

   In the context of disputes over the colony's direction, experiences of cultivation were forwarded as a symbol of the promise of the colony and proof
of the moral authority of its founders and leaders. There was nothing "curious" about the wheat, rye, barley, and oats that Poutrincourt brought back from
Acadia.[47] These were not rare plants valuable to collectors, nor were they of
any medicinal benefit beyond the calories that they could provide. These were
among the humblest plants that could be brought back from New France, but
in this Lescarbot suggested that Poutrincourt had echoed the Romans, who
had likewise translated such a harvest into symbols of triumph in their own
newly claimed regions, dating back to the foundations of Rome itself.[48] He

was, Lescarbot would reflect, akin to the "good father Noah, who after having made the most necessary agriculture in the sowing of wheat, put himself to planting the vine."[49]

Colonization increased French access to American naturalia, but increased access did not necessarily translate into greater knowledge about North America. French naturalists such as the Provence-based Nicolas-Claude Fabri de Peiresc were able to personally observe and experience North American animals such as the caribou, the hummingbird, and a horseshoe crab that had been brought back by early Acadian settlers such as de Monts.[50] These North American animals joined collections that also included cultural artifacts such as canoes, bows and arrows, and the aboriginal weapon known to the French as the *casse-tête*.[51] They were valued, like other objects with which they were stored and compared, less for their contribution to furthering knowledge about newly claimed or discovered regions of the world than for their novelty and rarity.[52] The arrival and dissemination of these American plants were therefore part of a much larger culture of curiosity that animated a diverse array of intellectual and commercial activity in the seventeenth century and that was at least equally concerned with geographical breadth as it was precision.[53]

Considerable geographical uncertainty limited the impact of flora from northeastern North America. Within scientific genres, the significance of locality remained a debated subject.[54] Terms used to designate regions of North America such as Canada remained unmoored to specific locations.[55] Early gardens such as Peiresc's blended flora from the world over. In a 1630 letter, for example, Peiresc described his garden in the southeast of France as home to "several curious pieces come from the Indies and from Canada and from elsewhere." Alongside "an orchard of fruit trees where I have more than sixty sort of excellent European apples," and hyacinths, he noted that his "vine of Canada" had "covered entire houses in three or four years."[56] The systematic study of these plants was further limited by their irregular arrival and frequently confused provenance. Peiresc and other collectors could touch, taste, smell, and observe American squash, grapes, and strawberries, but we have little record of where these originated from, nor do we have much information about the identities behind the many hands that would have been needed to transplant them in Peiresc's southern garden.[57] In hindsight it is possible to see the stirrings of a botanical science with a particular attention to morphological and geographical specificity, but that was not the goal of collectors such as Peiresc.

In Paris, it was the efforts of Guy de la Brosse, Jean Robin, and Robin's son Vespasien in the late sixteenth and early seventeenth centuries that established a sizable presence of American plants in the city's gardens. Both Robins and de la Brosse presided over gardens associated with royal authority and the city's medical establishments, Jean and Vespasien Robin as gardeners for Henri IV and the city's medical faculty and de la Brosse as physician to Louis XIII and first director of the Jardin du Roi after it was founded in 1635.[58] They directed increasingly sophisticated and well-funded gardens that helped the crown establish cultural authority through the skilled display of exotic plants.[59] Jean Robin, a surgeon by training, had been hired to create a garden for the faculty of medicine in Paris in 1597.[60] His son Vespasien collected plants throughout Europe and, where they had already been transplanted in European gardens, from the Americas and Asia.[61] Vespasien was later hired as a botanical demonstrator at the Jardin du Roi founded by de la Brosse.[62] In a 1641 catalogue of the plants that grew in his garden, de La Brosse credited Vespasien with introducing many of the foreign plants, as well as with maintaining the networks through which new seeds and specimens arrived.[63]

Vespasien and his father were only two among the many Paris-based gardeners who introduced foreign flora into the French capital in the sixteenth and early seventeenth centuries.[64] Published catalogues of their gardens allow us to trace the growth in the presence of American plants numerically, even if the source of many of these specimens remains unknowable. In 1601, at least three of the almost 1,300 plants that Robin listed in his *Catalogus stirpium tam indigenarum quam exoticarum quae Lutetiae coluntur* were identifiably American: an *arbor vitae* that was first brought back by Jacques Cartier, a *Christophoriana* that Robin also supplied to the English herbalist John Gerard, and the *Aconitum racemosum sive Christophoriana* that soon became better known simply as snakeroot.[65] If only a few were positively North American, many were rare and came to Jean through extensive networks that he cultivated with other collectors and that connected him with plants that were arriving from French, Spanish, Dutch, and English colonies in the Americas.[66] Within the first decades of the seventeenth century, he built a sizable collection that would later form the basis of the Jardin du Roi.[67]

Yet it would be a mistake to assume that the American origins of these plants were prized or that their cultivation supported conversations about colonialism in the places from which they had come. Jean Robin's association with *Robinia pseudoacacia*, also known as the Black Locust, for example, reveals the relative unimportance of origin to these collections. Jean is frequently acknowledged as having introduced the plant into Paris—and more specifically into the Left Bank garden of the Paris medical faculty near Notre Dame Cathedral—in 1601, where it grows to this day.[68] His son Vespasien planted another long-lived example from the seeds of this first *R. pseudoacacia* in 1636 in the Jardin du Roi where he worked as a botanical demonstrator and arborist.[69] The tree soon spread roots beyond the confines of Paris, although the eighteenth-century amateur botanist Joseph-Pierre Buc'hoz wrote that early experiments with planting the tree along French rural *allées* failed; unfortunately, he wrote, the branches broke "easily in the lightest wind," although parts of the tree were eventually used both medicinally and for woodworking.[70] Over a century after the first example of the tree was growing in Paris, Carolus Linnaeus attached Robin's name to what by then was increasingly known as a false acacia.[71]

Yet the exact origins of the tree planted in 1601 remain unknown and disputed to this day, at least in part because the first name provided—*Acacia Americana Robini*—failed to identify either a specific region of origin or a vector of its arrival. Some scholars have suggested that the tree was first acquired much later than 1601 from the English naturalist John Tradescant, who had acquired it for himself from Virginia, either from his son John the younger (who traveled to Virginia on a botanizing trip) or from correspondents who had settled in the colony.[72] Although Champlain might seem to be a possible source, his trip to Mexico and the Caribbean left him far to the south of the tree's natural range, and his explorations of Acadia and the Saint Lawrence Valley were both too late and too far north.[73] In this respect, the uncertain origin of *R. pseudoacacia* makes it representative of many of the early plants that crossed the Atlantic from North America.

When Jacques Philippe Cornut produced the first written description and visual image of *Robinia pseudoacacia* in his 1635 *Canadensium plantarum* he called it *Acacia Americana Robini*; but he used the terms "Canada" to represent a far larger region than the present-day country and "America" as a fluid geographical marker more akin to how we today use the term "Americas" (Figure 5).[74] The text marked a transition between the humanist herbals of the sixteenth century and the regional floras of the seventeenth and the eighteenth. The passages that described the plants and the copper-plate images that represented them were not the product of the circumscribed field trips that would soon come to define regional floras but were instead an effort to expand the geographical coverage of classical botanical authorities.[75] If he and others of this community of Paris-based natural historians therefore continued to confuse

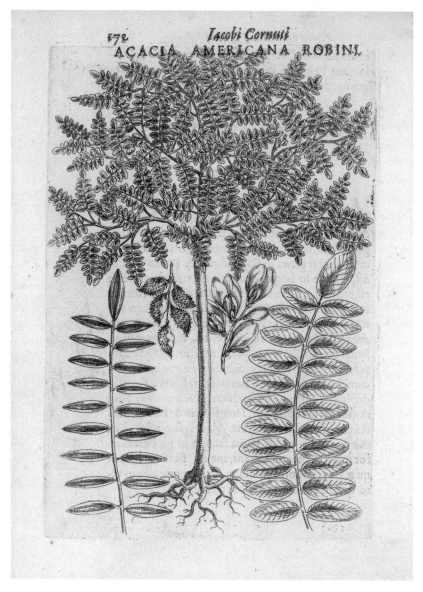

Figure 5. Jacques-Philippe Cornut, "Acacia Americana Robini,"
Canadensium plantarum, 1635. Courtesy of the John Carter Brown Library
at Brown University, Providence, Rhode Island.

the specific geographical origins of the plants they described, they surely can be forgiven; this was not the task that they took up for themselves or their science.

*   *   *

Successful calls to cultivate New France required methods of communication that could simultaneously describe New France for French readers and diagnose its deficiencies. The specimens in Parisian gardens could not accomplish this. If we were to look for an author who can offer an image of a typical French traveler of this style in the interior of North America, we might easily pick Jean de Brébeuf. Arriving initially in New France in 1625, he is best remembered for his exploration of the Great Lakes and his residence among the Wendat people in what is now Ontario. Brébeuf traveled widely throughout Wendat territory and was an early and authoritative source of information about the region that would become the *pays d'en haut*. About his life among the Wendat he wrote:

> We live on the shore of a great Lake, which affords as good fish as I have ever seen or eaten in France; true, as I have said, we do not ordinarily procure them, and still less do we get meat, which is even more rarely seen here. Fruits even, according to the season, provided the year be somewhat favorable, are not lacking to us; strawberries, raspberries, and blackberries are to be found in almost incredible quantities. We gather plenty of grapes, which are fairly good; the squashes last sometimes four and five months, and are so abundant that they are to be had almost for nothing, and so good that, on being cooked in the ashes, they are eaten as apples are in France. Consequently, to tell the truth, as regards provisions, the change from France is not very great; the only grain of the Country is a sufficient nourishment, when one is somewhat accustomed to it. The *Sauvages* prepare it in more than twenty ways and yet employ only fire and water; it is true that the best sauce is that which it carries with it.[76]

Like the *Relations* more broadly, this is a complicated text that transitions cleanly between a single authorial voice and a "we" that spoke to both his fellow travelers and his readers. Brébeuf acknowledged both the limits of his own experience ("we do not ordinarily procure them") and his reliance upon indigenous knowledge and labor.[77] He nonetheless confidently named types of edible plants that he clearly knew intimately and expected his readers to know as well.

We can see then how such narratives were a deceptively simple formal strategy for describing newly discovered places and evoking the promise of colonialism to draw out latent possibilities in places and peoples. Even as the decades passed and the geographical reach of missionary authors increased, travel narratives presented their authors with an effective means to translate their experiences in a manner meant both to entice support and to assuage any concerns about the illegibility of new flora and environments. Writing about a Mascouten village in the western Great Lakes to which he had traveled in 1673, for example, Jacques Marquette wrote that "I took pleasure in observing the situation of this village. It is beautiful and very pleasing; For, from an Eminence upon which it is placed, one beholds on every side prairies, extending farther than the eye can see, interspersed with groves or with lofty trees. The soil is very fertile, and yields much Indian corn. The *sauvages* gather quantities of plums and grapes, wherewith much wine could be made, if desired."[78] Like Brébeuf, Marquette acknowledged aboriginal presence but confidently anticipated the expansion of French colonialism that would enrich the agricultural and ecological productivity of the region. Marquette assured his readers that familiar plants dotted the landscape, but he also introduced his own aesthetic judgments to provide an assessment of the innate beauty of the region and possibilities for French improvement.

The centrality of cultivation to the propagandistic quality of these writings is marked. In the 1632 edition of his *Voyages*, for example, Samuel de Champlain promised Richelieu an account of "lands no less than four times the size of France, as well as the progress in the conversion of the *sauvages*, the clearings of some of these lands whereby you will perceive that in no respect are they less fertile than that of France; and finally the settlements and forts that have been built there in the name of France."[79] It was a project that he summarized as aiming to "restore and retain possession of this New Land by the settlements and colonies which will be found necessary there."[80] Discussions of a *sauvage* country implied that conditions that might be claimed to be defining features of these newly explored and settled regions of Acadia and Québec were in fact remediable defects. The ambition of this era was perhaps best captured by the Jesuit Pierre Biard, however, who wrote that "there is no reason why the soil should not be equally fertile, if the cultivation of the plains were long continued upon to lands, and if it were not for the dense shades of the almost unbroken forests."[81] New France offered nothing less, he wrote, than "another France . . . to be cultivated."[82]

Communicating American flora in print was therefore not an exercise in abstraction. Descriptive detail in these accounts gathered where authors allowed themselves to inhabit a place or to imagine it in the not-too-distant colonial future. The organization of these accounts was therefore both spatial and temporal, offering experience of linear itineraries and selected sites that were transmuted into representative samples of larger American environments.[83] Scholars of the literatures of encounter in French North America have often emphasized the distinction between colonial and missionary texts of this early period, yet the shared reliance on chronologically organized narratives produced similar descriptions of early American places.[84] They were accounts that often provided brief descriptions, offering the perspective of a traveler who could only catch glimpses of complex ecologies from canoes or on foot and who often relied on guides as they traveled into the heart of the continent. These were environments that were peopled, either by living indigenous communities or by the specter of future colonial development. Authors explicitly presented spaces in which they had dwelled; they foregrounded their own experience and knowledge produced through their physical presence in American cultural and ecological settings.[85] Embedded in descriptions of unfamiliar peoples and landscapes, narratives such as these highlighted the presence of recognizable French plants such as oak, grapevine, strawberries, and grains that anchored readers and travelers alike, continuing to offer promises of an essential familiarity behind apparent difference.

In effect, plants and American environments continued to be known as they were lived, and the immersive perspective of early authors such as Champlain, Sagard, and Le Jeune continued throughout the century. North American plants could be identified by functional roles they shared with European counterparts, an essence defined, at least in part, by how French communities could live with a plant. Describing Acadia in his 1672 *Histoire naturelle*, the colonial promoter and landowner Nicolas Denys wrote, "There are also pine for making planks, good for making decks, and fir for ornaments . . . pine, little spruce and fir are also found in the forests of this country which serve for tar the qualities of which I have already spoken."[86] If the American wilderness was perceived through a lens that favored extractive enterprise and the transformation of botanical resources into commodities, seventeenth-century texts from French North America suggest that colonists and missionaries also looked at North American flora with an eye to transplanting

European ecological relationships into new soils. More than a mercantilist gaze, this was an understanding of botanical identity that saw latent or potential utility as a constitutive facet of a plant's identity.[87] The adoption of the natural historical genre allowed even greater flexibility to include knowledge acquired from decades of colonial experience. It meant that even the imposing forests of New France were seen to support the efforts by colonial authors who understood that this flora would be called upon through cultivation of French colonial spaces and lives.[88]

Plants in print—those that arrived as descriptions in travel narratives, histories, natural histories, and personal correspondence—therefore contributed far more to an emerging awareness of Acadian and Laurentian places and peoples than the specimens that grew in French gardens. Sagard wrote in 1632 that some of his fellow missionaries had brought "some *Martagons*" to France, "with some cardinal plants as rare flowers, but they did not profit there, nor did they reach their perfection, as they do in their own climate and native soil."[89] He juxtaposed this anecdote with his description of the landscapes of the early seventeenth-century Saint Lawrence Valley, suggesting that he hoped instead to offer his French readers a virtual experience of the plant's "native soil" through his text.[90] Not only were written descriptions better able to survive transit and inform a broader audience in France, they did the additional work of bringing whole environments to life, rather than single specimens.

Like Poutrincourt's presentation of wheat, authors such as Lescarbot and Champlain took presentations of American flora—both native and introduced—as opportunities to valorize experiential ways of knowing these newly colonized places that made them legible and familiar to their audiences. Travel narratives offered opportunities to convey personal experience of American places. By the time that missionary *relations* and the *récits de voyage* of explorers presented New France to European readers, travel narratives were well established as a privileged genre for carrying experiences of the new worlds Europeans discovered around the globe back to reading publics in Europe.[91] In the representation of New France, the travel narrative provided authors with an opportunity to meditate on the most prosaic fixtures of American landscapes, establishing familiarity and engendering confidence in the success of French colonialism to make them more familiar still.[92] Even as authors began to call their accounts natural histories by midcentury, a strong authorial presence remained, along with considerable reliance on narratives that warranted descriptions as the product of firsthand experience. Narratives allowed authors to seamlessly move between empirical observations of real environments and imagine a not-too-distant promised future.[93] The movement between these two tenses both legitimated colonial authors as experts and imaginatively engaged readers in the cultivation of an empire in northeastern North America.

The genres favored by colonial authors were the most capacious, and accounts of early American environments featured a variety of written forms.[94] If the forms were fluid, however, the function remained consistent: to capture not only the experience of a New World but the affinities that connected New France and Old and that transcended the physical distance of the ocean that separated them. Textual features such as lists, indigenous-language dictionaries, and specific sections that dealt with natural historical subjects hint at the variety of information that could be contained within texts that modestly claimed to be simple accounts of circumscribed travel.[95] For example, the dictionary included in Sagard's *Grand Voyage*, like the chapter devoted to plants, offered Native terms for plants seemingly abstracted from their local cultural or ecological context alongside vocabulary related to the consumption of tobacco and other foods, farming, and the medicinal use of local plants.[96] In this, Sagard joined other authors such as Lescarbot in a willingness to enrich their own narratives with features common to other genres. The Acadian lawyer's *Histoire*, for example, combined a first-person narrative of early maritime colonization with an anthology of previous and contemporary French efforts in the region. In the process, his text became a palimpsest of forms and narrative techniques that converged to make New France legible as a colonial space.[97]

As part of a broader effort to centralize his authority, in 1663 Louis XIV took direct control of New France and reorganized its government. As the king's interest in France's colonial possessions grew, Pierre Boucher, Nicolas Denys, Louis Nicolas, and other authors expanded the effort to explain New France to multiple and more popular audiences across the Atlantic. The justifications for these efforts varied. Some, such as Louis Nicolas, had left New France behind and sought to use knowledge acquired there to build a life and reputation for themselves across the Atlantic.[98] Others, such as the landowners and promoters Nicolas Denys and Pierre Boucher, while likewise staking a claim to authority based on their considerable firsthand experience, claimed that they did so to counter false testimony that had degraded the image of New France and that had undermined the appeal of colonization there. Boucher claimed the goal of telling his readers "the truth with the greatest naïveté that is possible, and the briefest that I can."[99] Nicolas Denys likewise sought to "disabuse" his readers of pernicious false opinions that he himself had been

subject to before his arrival in Acadia.[100] In this they implicitly joined the efforts of their colonial predecessors, yet they did so in a genre that had thus far
had little role in the works of those authors who related New France. The
genre that Denys, Nicolas, and Boucher chose was the natural history.

Like travel narratives, the genre of the natural history was in flux during
this period.[101] The major impact of the natural historical texts that described
New France was to remove the chronological and linear focus that had defined earlier accounts. In his *Histoire veritable*, for example, Boucher organized
many of his chapters around specific kinds of life recognizable to seventeenth-
century authors, such as trees, animals, birds, and fish, but also included chapters devoted to particular regions such as Québec and others that addressed
indigenous peoples.[102] Thus even where the descriptions of plants seem haphazard and chaotic to the modern reader, for early modern audiences they assumed a familiarity with organizing categories such as trees, grasses, or bushes
common in contemporary botanical texts; their inclusion in texts about North
American flora argued for the existence of fundamental similarities between
North American and European plants.[103] Louis Nicolas, for example, divided
his descriptions of over two hundred plants from sections on fish, birds, mam-
mals, and aboriginal peoples and provided smaller sections that divided trees
from shrubs and grasses from fruits.[104] Other natural histories were similarly
organized.

At least some travel narrative authors similarly sought to escape, at least
temporarily, the linearity of their narratives to linger on a discussion of regions in an abstract language that could encompass spaces broader than they
might otherwise be able to include. Narratives such as Louis Hennepin's 1683
Description de la Louisiane, for example, seemed willing to blend formal elements of both narrative and natural history.[105] The first sections of the book
included information presented in a manner that would have been familiar
to readers of the *Relations* or Champlain's *Voyages*. His description of a Detroit "covered by forests, fruit trees such as walnut, chestnuts, prune trees, apple
trees, [and] *sauvage* vines, charged with grapes," was introduced alongside the
information that he and his fellow travelers were "fortunate enough to have
arrived at the entrance of Détroit on the tenth in the morning, the feast day
of Saint Lawrence."[106] These economical descriptions were complemented,
however, by an appended text titled *Les moeurs des sauvages* that also included
an introductory chapter on the "fertility of the country of the *Sauvages*."[107]

In place of an emphasis on the different genres in which authors such as
Champlain, Lescarbot, Boucher, and Denys wrote, we should instead empha-

size both a common epistemology and shared formal promiscuity. Lists appeared within narratives, and accounts of travel punctuated natural historical texts. Consistently, however, these authors rarely failed to establish their authority as firsthand observers who knew the places that they described through their own labors of cultivation and experimentation.

*   *   *

Authors such as Champlain, Lescarbot, and Biard established themselves as experts on American environments and made effective use of genres that privileged their own experience and knowledge. Champlain's various *Voyages* and Lescarbot's editions of the *Histoire de la Nouvelle-France* focused primary attention on the significant firsthand experience that they had acquired in the colony.[108] While both Biard's and Gabriel Sagard's accounts of their missions to American indigenous peoples were interested in the invisible and otherworldly, their narratives revolved around their explorations—around their actions and experiences.[109] Sagard promised, for example, that "I speak only of what I am assured."[110] Chronologically organized accounts emphasized the complexity of American environments and privileged the expertise of authors who had spent considerable time in New France.[111] Descriptions of the weather, for example, became opportunities both to relate empirically observed facts and to remind readers of the length and breadth of an author's experience. Foregrounding his experience of both Frances, Biard wrote, for example, that "I noticed once, that two February days, the 26th and 27th, were as beautiful, mild, and spring-like as are those in France about that time; nevertheless, the third day after, it snowed a little and the cold returned. Sometimes in summer the heat is as intolerable, or more so than it is in France; but it does not last long, and soon the sky begins to be overcast."[112] Champlain recounted successful experiments with spring and winter plantings, and Lescarbot reported that promoters had brought back some samples of Old World crops grown in Acadia to accompany his written account of the fecundity and—more important—the reliability and predictability of American climates.[113] Each emphasized the empirical foundation of their knowledge and used their own experienced bodies as a metric for their readers, who were instructed how to appreciate the similarities between France and New France.[114]

Narrators also claimed a role as guarantors of testimony collected from indigenous and colonial sources. In many of the early *Relations* from the 1630s, for instance, Jesuits such as Paul Le Jeune established themselves as both collectors of testimony and witnesses in their own right.[115] Le Jeune wrote in 1633, for example, that "on the 28th [of October], some French hunters, returning from the islands which are in the great St. Lawrence River, told us . . . that there were apples in those islands, very sweet but very small; and that they had eaten plums which would not be in any way inferior to our apricots in France if the trees were cultivated."[116] When Jacques Bruyas wrote in a letter from the mission to the Haudenosaunee at Saint François-Xavier in 1668 that "apple, plum, and chestnut trees are seen here," he provided little sense of when or by whom they were seen.[117] These movements between firsthand observation and gathered testimony served to foreground the author as expert and broaden the field of observation to verify observed and expected facts.[118] Colonial authorities such as Champlain similarly translated their social privilege into epistemological authority that enabled him to speak of and on behalf of New France.[119] A counterpart to his dual role in New France as a civil authority and an explorer, Champlain switched readily between his own experiences and those of others over whom he governed in his various *Voyages*.[120] We can read as easily, for example, about his winter experiences among the Wendat or his explorations of the Ottawa River as we can his summaries of the experiences of colonists such as Louis Hébert who are otherwise silent.[121] Jesuit and colonial authors communicated experience of American flora as a composite of multiple experiences by multiple authors and witnesses.[122]

Whether forwarding gathered testimony or relating their own considerable experience, the authors of natural histories and travel accounts to New France such as Champlain and Jesuit missionaries favored narratives that resisted reducing botanical knowledge to the description of plant morphology. Visual observation was privileged, but sight as it was increasingly used within contemporary European science was rarely deemed sufficient in and of itself. Renaissance and early modern natural history focused relentlessly on the visible characteristics of plants, and the growing use of textual descriptions and dried herbarium specimens in lieu of direct experience of living plants marginalized descriptions of what we might call their ecological contexts. In contrast, North America–based authors continued to situate novel flora in narratives that immersed their readers in complex and irreducible ecosystems.[123] Both authors of early seventeenth-century travel narratives and authors of later natural histories such as Nicolas Denys and Louis Nicolas clearly appreciated the visible qualities of American flora, yet they rarely failed to also reference the tastes, smells, and aboriginal uses of new plants and foods. Even as empirical experience was central to French accounts, then, the field of experience and the types of knowledge presented remained self-consciously broad.

The result was that where modern readers might expect botanical descriptions that focus on morphology, other qualifications were frequently interwoven into narratives that drew upon multiple senses and that blended ethnographic and botanical observations. Take, for instance, the following account of a new plant described by Louis Nicolas in his *Histoire naturelle*, written around 1675. Nicolas named the plant simply "another black fruit."[124] This, in itself, was neither out of the ordinary for natural historical texts nor particularly informative. When he later added that Europeans could not accustom themselves to its taste, however, he took what might seem to be a strange rhetorical turn, writing that "it seems that these strange people have an aversion to everything that we like, and prize everything that we despise; they cannot bear our best smells, and say that they smell bad."[125] Investigations of plant life invited commentary on local cultures, and vice versa. This is not to suggest that missionary authors elided natural and cultural descriptions, categories that for missionaries and their readers meant little at the time.[126] Rather, it hinted at the belief in a complex web of relations between people and place. As Jesuits worked to convert souls and colonists worked to transform place, they became aware of the importance of flora as emblematic of broader features of the North American natural and cultural landscape.

Narratives that recounted experience of peopled landscapes made the colonialist intent of natural description particularly clear; French texts could provide both experience of new flora and judgments about the inadequacy of indigenous ecological knowledge. In 1639, for example, as the Jesuit Paul Le Jeune sought to provide his readers with insights into the "superstitions" and "customs" of the Algonquian speakers whom the missionaries were actively trying to settle at Québec in agricultural communities, he took what might seem to us a strange turn. As he proceeded "first, as to what concerns their belief," he soon wrote as much about fruit as he did about peoples. He explained:

> Some of them imagine a Paradise abounding in blueberries; these are little blue fruits, the berries of which are as large as the largest grapes. I have not seen any of them in France. They have a tolerably good flavor, and for this reason the souls like them very much. Others say that the souls do nothing but dance after their departure from this life: there are some who admit the transmigration of souls, as Pythagoras did; and the majority of them imagine that the

soul is insensible after it has left the body: as a general thing, all
believe that it is immortal. . . . In fact, I have heard some of them
assert that they have no souls; they hear people talk about these
attendant forms, and sometimes persuade themselves that they
possess them,—the Devil employing their imagination and their
passions, or their melancholy, to bring about some results that
appear to them extraordinary.[127]

Le Jeune's description of the blueberry functioned as a means to explore and
explain aboriginal conceptions of the soul to his European readers and hinted
at the possibility of diabolical influence in indigenous religions.

First-person narratives permitted moving seamlessly between the botanical and ethnographic description that enabled authors to both catalogue
American flora and diagnose it as deficient.[128] Writing from Wendat territory
in 1653, for instance, Bressani declared that "there are some wild vines, but in
small quantity, nor are they esteemed by the Barbarians themselves; but they
do esteem highly a certain fruit of violet color, the size of a juniper berry which
I have never seen in these countries. I have also seen, once, a plant similar to the
Melon of India, with fruit the size of a small lime."[129] Writing from the mission
at Kahnawake, Jacques Frémin similarly wrote: "And besides the grapes,
plums, apples, and other fruits, which would be fairly good if the *Sauvages*
had patience to let them ripen, there also grows on the prairies a kind of lime
resembling that of France."[130] Jesuits presented cultural and natural environments that were defined by aboriginal ecological practice. Even where authors
such as Frémin and Bressani added their own observations, they were assembled from personal experiences that were couched in wider discussions of
aboriginal ecological lives.

Subjective experiences therefore became the primary registers for explaining these new foods, plants, and animals. Encounters with indigenous food
provoked few of the anxieties that wracked English and Spanish explorers who
feared that changes to their humoral complexion would result from consum-
ing foreign foodstuffs.[131] The primary challenge was the disgust and discomfort that were prominent in written accounts of Native foodways. Sagard's
description of a dish of fermented corn among the Wendat, for example,
evoked a plate "very stinky and more rotten than even the gutters."[132] Among
the Innu, Paul Le Jeune blurred natural history and a description of indigenous foods when he gave an account of "the meats and other dishes which
the *sauvages* eat, their seasoning, and their drinks." He noted that

among their terrestrial animals they have the Elk, which is here generally called the Moose; Castors, which the English call Beavers; Caribou by some called the Wild ass; they also have Bears, Badgers, Porcupines, Foxes, Hares, Whistler or Nightingale,—this is an animal larger than a Hare; they eat also Martens, and three kinds of Squirrels. As to birds, they have Bustards, white and gray Geese, several species of Ducks, Teals, Ospreys and several kinds of Divers. These are all river birds. They also catch Partridges or gray Hazel-hens, Woodcocks and Snipe of many kinds, Turtle doves, etc. As to Fish, they catch, in the season, different kinds of Salmon, Seals, Pike, Carp, and Sturgeon of various sorts; Whitefish, Gold-fish, Barbels, Eels, Lampreys, Smelt, Turtles, and others. They eat, besides some small ground fruits, such as raspberries, blueberries, strawberries, nuts which have very little meat, hazelnuts, wild apples sweeter than those of France, but much smaller; cherries, of which the flesh and pit together are not larger than the pit of the Bigarreau cherry in France. They have also other small Wild fruits of different kinds, in some places Wild Grapes; in short, all the fruits they have (except strawberries and raspberries, which they have in abundance) are not worth one single species of the most ordinary fruits of Europe.[133]

Details of subjective experiences such as tasting Native foods reveal the colonialist intent of these otherwise descriptive accounts. Whether travel narrative or natural history, description was not an end in and of itself. Instead, authors such as Le Jeune seized upon these moments to invite readers to understand the need for French intervention in cultural and natural landscapes that were deemed lacking.

Authors who based their descriptions on their own experiences positioned their bodies as instruments in the production of natural knowledge and their own subjective tastes as the metric for comparisons between New France and Old. In 1640, for example, Pierre-Joseph-Marie Chaumonot described *sagamité*, a porridge that the Wendat made with corn, as he wrote that "the whole apparatus of our kitchen and of our refectory consists of a great wooden dish, full of *sagamité*, whereto I see nothing more similar than the paste which is used in covering walls. Thirst hardly annoys us,—either because we never use salt, or because our food is always very liquid. As for me, since I have been here, I have not drunk in all a glass of water, although it is now eight months since

I arrived."[134] A letter from François du Peron to his brother from the Wendat village of Ossossanë written a year earlier similarly explained that "one does not have undisturbed rest here, as in France; all our Fathers and domestics, except one or two, I being of the number, rise four or five times every night . . . the food here causes this."[135] In this way, du Peron's digestion became an opportunity to highlight deficient indigenous relationships with American nature; colonial bodies became an essential and authoritative mediator of both American environments and authorized judgment of aboriginal ecological practice.

Narratives equally promised that *sauvage* plants that might differ subtly from their French counterparts could nonetheless be counted upon to support French lives. While the use of a name such as "oak," "cherry," "vine," or "lemon" certainly implied the existence of specific morphological features expected by French audiences and colonists, these familiar French names also implied a set of potential uses and ways of living with the plant and its products. On both sides of the Atlantic, North American plants became knowable through lived experience of working with them, and their possible uses and incorporations into French ecological and domestic regimes figured prominently in early written accounts. Experience with American crops demonstrated that they could also be understood through practice—through experience of cultivation, harvest, cooking, and digestion. Where corn was integrated into French fields and lives, for example, it was because it was valuable as a substitute for other grains. In Nicolas de Ville's *Histoire des plantes de l'Europe et des plus usitées qui viennent d'Asie, d'Afrique et de l'Amérique* he wrote, for example, that "the flour is white . . . but thicker and more viscous than that of wheat; it is less easily digestible. The peasants make a porridge of it with butter and cheese which is agreeable enough, even if heavy on the stomach. The flour is excellent for plasters which ripen. The juice of the leaves is good for inflammations and erysipelas."[136] Where botanical science relied upon morphology and other visual cues to provide differentiation, narrative and a dwelling perspective could promise familiarity far more effectively.

Only a few decades later in 1709, Louis Liger confidently assumed a widespread knowledge and experience of the plant when he wrote that "the Turkish wheat, otherwise known as Indian wheat, is known well enough, such that there is no need to describe it."[137] After detailing the method and timing necessary to plant the crop, Liger continued to situate the plant within a French geographical and social setting.

It is not difficult to acquire Turkish wheat to sow, because it is very common in Burgundy, in Franche-Comté and in Bresse where a lot of it is cultivated, its usages . . . are very advantageous, the grain is milled and the flour is used to make bread of which almost all the laborers of these regions feed their families during the entire year. The flour is also used to make *beignets*, *galettes*, tarts seasoned with dairy products, and a type of porridge that they call *Gaude*, that they make like rice or millet; this serves as breakfast for everyone in the house, it is for this reason that from the morning on a pot is put in front of the fire, then when the *Gaude* is cooked, each can take a full bowl, which is enough to fill the stomach to capacity.[138]

While French authors in North America continued to discuss the aboriginal custom of planting corn in mixed fields with beans and squash, they were nonetheless equally informed by a growing French confidence in the ability to know American flora through its function in French lives. Cultivating American flora became a central way of knowing New France and therefore figured prominently as both an act and a metaphor of France's colonial project in North America.

*   *   *

An entry in the *Journal des Jésuites* for 1641–42, a document that emerged from the injunction of Jesuits to document their day-to-day lives, made the georgic view of French colonialism explicit. It explained that "the grains have been very beautiful. Some *habitants* now harvest more than they need to feed their family and their animals that are doing well in this country here. The time will come when all will have the same. *Labor improbus omnia vincit*."[139] The Latin quote is a passage from Virgil's *Georgics* that, in a recent edition, has been translated as "hard work prevailed" but that has commonly been read as meaning that "labor conquers everything" or that "toil and the pinch of need drove men on, with the result that they succeeded in defeating the obstacles before them."[140]

Virgil's *Georgics* was not entirely a surprising source for the anonymous author of this passage in the *Journal des Jésuites*.[141] Virgil, alongside other classical authorities on agriculture and gardening such as Columella, Cato, Varro, and Palladius inspired renaissance horticulturalists immensely in editions and translations that were published repeatedly throughout Europe.[142] The *Georgics* was particularly prominent throughout the ancient Mediterranean world, and it was widely read in early modern Europe as the georgic, alongside the pastoral to which it is often opposed, became an important mode within which relationships to the land were read and recast.[143] Virgil was read by Jesuits during their formation alongside other Greco-Roman authors such as Homer and had an outsized influence on the narrative structure of, among others, the Jesuit *Relations* through which so much of the early history of New France is known.[144] Virgil had been rehabilitated by humanists who, by the end of the sixteenth century, were comfortable recontextualizing his poetry within a Christian conception of agricultural labor.[145]

This passage is one Latin quote among many in the writing of the Jesuits who came to New France in the seventeenth century, but it is particularly illuminating as it reveals the extent to which considerations of agriculture and the environment—of cultivation—could wed together discussions across historical, linguistic, and disciplinary divides. The passage is a direct reference, of course, to the classical heritage of colonialism France was grappling with and seeking to work through in its own overseas expansion. The *Georgics* were written during a period of crisis in Roman agriculture that saw the widespread use of slavery, the expansion of the holdings of large landowners, and efforts to centralize imperial authority. The title might translate from a Greek term meaning "things related to farming," but the book was very much understood to explore the entanglement of human and natural environments.[146]

Above all, however, this passage testifies to the optimism of early modern colonialism in what is now Canada and the faith that agricultural labor was both a divine command and divinely ordained to succeed. The task that this Jesuit author and the other Europeans who had come to New France with him had taken upon themselves was enormous. They laid claim to nothing less than the rehabilitation of an errant continent that had been left to become *sauvage*. *Labor improbus omnia vincit* suggests where this optimism came from and the tenacity of the idea. Blending classical example with faith in providential design, they could look out onto the strangely familiar landscapes of northeastern North America and truly see a New France waiting to be cultivated.

CHAPTER 3

Cultivating Soils and Souls

When the Jesuit superior Paul Le Jeune planned for the salvation of the Innu peoples with whom he had only recently wintered, he conceptualized their conversion in deeply ecological terms.[1] "If they are sedentary, and if they cultivate the land," he promised, "they will not die of hunger, as often happens to them in their wanderings; we shall be able to instruct them easily." Wild forests, depleted animal populations, and hunger-wracked indigenous bodies were material manifestations of a broader spiritual malaise and a failure to properly manage the natural and spiritual abundance that Christianity would make possible. The beaver, too, would be saved as part of a broader reorganization of indigenous lives, economies, and environments. While "these animals are more prolific than our sheep in France, the females bearing as many as five or six every year . . . [t]here is danger that they will finally exterminate the species in this Region, as has happened among the Hurons, who have not a single Beaver, going elsewhere to buy the skins they bring to the storehouse of these Gentlemen," he opined. Nonetheless, as the Innu adjusted to settled agriculture, "in the course of time, each family of our Montaignais, if they become located, will take its own territory for hunting, without following in the tracks of its neighbors; besides, we will counsel them not to kill any but the males, and of those only such as are large." The conversion of nomadic people to agricultural lives was therefore one piece of a broader ecological transformation, for "if they act upon this advice, they will have Beaver meat and skins in the greatest abundance."[2]

Missionaries frequently engaged horticultural thought as a metaphor through which to explain the process of conversion.[3] In 1616, the Jesuit Pierre Biard had made sense of the communities with whom he lived by looking to the prophet Isaiah, who had written within "a fitting and appropriate comparison of a great orchard or garden, wild and uncultivated."[4] Indigenous communities, Biard wrote, "through the progress and experience of centuries, ought to have come to some perfection in the arts, sciences and philosophy" but were nonetheless "like a great field of stunted and ill-begotten wild plants, a people which ought to have produced abundant fruits in philosophy, government, customs, and conveniences of life." They were people who "ought to be already prepared for the completeness of the Holy Gospel, to be received in the house of God."[5] This elaborate botanical metaphor claimed that the current state of the Mi'kmaq to whom Biard proselytized was neither natural nor sustainable and invited the intervention of those like him who were willing to travel to this New France in North America to work souls lost to neglect so that the Christianity that was their salvation might soon find more fertile ground. Biard asked God, "Wilt thou not look upon this poor wilderness with a favoring eye? Kind and pious husbandman, so act that the prophecy which follows may be fulfilled upon us and in our time."[6] The project of conversion was made legible through a horticultural metaphor in which the act of cultivation revealed the true nature of *sauvage* peoples.

In other accounts, both near contemporaries to those of Biard and those written throughout the seventeenth century, indigenous peoples were similarly cast as wild vines, untended crops, and fallow fields. In 1640, Le Jeune explained that Jesuits and other missionaries who had come to New France labored to "cultivate these young plants, and to render them worthy of the garden of the Church, that they may be some day transplanted into the holy gardens of Paradise."[7] Within this horticultural idiom, indigenous peoples who converted were "wild plants of these countries, transplanted into the Church of God."[8] As he explained in the following year's *Relation*, "The word of the Gospel germinates when the holy Ghost wills to render it fruitful; it is for us merely to sow it with fidelity, and await heaven's moments."[9] As a metaphor, horticulture mapped missionary activity and the reception of Christianity onto a teleological account of plant development and cultivation; missionaries became gardeners and indigenous conversion the natural product of their tending to a *sauvage* people.

Yet Le Jeune's hope for the beavers of the Saint Lawrence Valley reveals that cultivation became much more than a framing metaphor for the colonial process as it drove the missionaries who came to New France in the seventeenth century to look for answers about indigenous souls in American soils.[10] Missionaries worked diligently to categorize indigenous ecological practices, the character of their foods, and the other relations with the natural worlds that sustained their communities. At sites such as the reduction of Sillery, these missionaries imagined that some of the worst environments of the colony could be saved in tandem with those nations who—"errant" people seen as dependent on natural charity—were most at risk. Cultivation, in these places, became less a metaphor than a mission itself that took down forests, threw up fields, and produced crops of European and indigenous plants.

Cultivation transformed colonial landscapes but it also pushed missionaries to engage with indigenous peoples and environments. The gardens, fields, and forests of New France's mission to Algonquian and Iroquoian peoples also became places of encounter. It was in these environments that missionaries were exposed to other ways of inhabiting American ecosystems and where they began to appreciate that the diverse cultures of indigenous North America possessed unique ecological knowledge. Missionaries learned the complex ties that bound their would-be flock to local places and gained glimpses of ways of belonging to these environments that elided the neat distinctions between *sauvage* and cultivated that had brought them to these places and inspired their practice. Cultivation naturalized French colonialism, but in encouraging the empirical study of American environments it also opened opportunities to encounter the knowledge of indigenous peoples and the agency of indigenous flora.

*   *   *

Civilization and conversion were closely twinned concepts in the strategies of the Franciscan and Jesuit missionaries who came to New France in the seventeenth century.[11] "Convert" was both noun and verb in Récollet and Jesuit accounts of the period; "civilization" cohered around identifiable habits of dress, food, religion, and daily life but lacked a precise definition.[12] Together they spoke to the internal and external transformation of indigenous lives, but their exact relationship remained elusive.[13] In their capaciousness, both provided useful ambitions and suggested a loosely defined set of practices with which missionaries could turn the ethnically, linguistically, and culturally diverse indigenous communities that they encountered after their arrival in this New France toward Christianity.[14]

A horticultural discourse that equated "civilization," "conversion," and related terms such as "frenchification" (*francisation*) with careful, directed cultivation grounded these abstract ideas and encouraged missionaries to study the material lives of their indigenous hosts. These missionaries drew a close

connection between indigenous spirituality and the social and ecological practices that sustained Native communities.[15] This was a natural outgrowth of the
"environmentalism" of early missionaries, who saw a close relationship between
how indigenous peoples lived and their spiritual lives.[16] The conviction that
relationships with the natural world could offer insight into the inner lives of
aboriginal cultures drove missionary authors to study their environments closely
and encouraged them to explain the flora, fauna, and places of New France
as influential actors in the colonial drama unfolding along the Saint Lawrence River and throughout the Great Lakes region.

When the historian George Stanley wrote that the Saint Lawrence Valley
was where "Le Jeune, the Jesuit Superior, sent forth his men to cultivate the
mission fields of Canada in the early years of the 17th century," he echoed
the language of the missionaries themselves.[17] Le Jeune, the Récollet Joseph
Le Caron, and others who described the early evangelization of Algonquian
and Iroquoian peoples conceived of their labor within a horticultural discourse
that equated their activities with gardeners, vintners, and farmers and that
foregrounded both real and metaphorical plants.[18] Jesuit missionaries, for
example, often wrote about the introduction of Christianity in a language that
evoked the intentional introduction and cultivation of a European plant.[19] In
1636, Jean de Brébeuf wrote, "You remember that plant, named 'the fear of
God,' with which it is said our Fathers at the beginning of our Society charmed
away the spirit of impurity; it does not grow in the land of the Hurons, but
it falls there abundantly from Heaven, if one has but a little care to cultivate
that which he brings here."[20] Only a few pages earlier, Brébeuf had commented
on "strawberries, raspberries, and blackberries" that "are to be found in almost incredible quantities."[21] A few pages later, he described the central role
of pumpkins in the transition of Wendat dead to the "Village of souls."[22]
Brébeuf was not alone in sowing these references throughout his text. In each
missionary *Relation*, the ubiquity of references to real and metaphorical plants,
to symbolic harvests and real feasts, and to imagined vineyards and richly tex-
tured indigenous horticultural spaces focused attention on the botanical as a
central avenue for learning about and living in New France.[23]

Within a decade of arriving in New France, missionaries such as the Jesuit
Pierre Biard and the Récollet Le Caron came to understand that transforming indigenous relationships with their environment was an essential pre-
requisite for the sort of religious conversion that they expected.[24] Le Caron
explained in a letter later reprinted by his fellow Récollet Chrestien Le Clercq
that "to convert them it is necessary to familiarize and settle them among us.

And now, when the colony has not multiplied and expanded when they have passed a month with us they need to go to war, to hunt, or to fish to find what they require to live; this debauches them, it is necessary therefore to fix them, and push them to clear and cultivate the soil, to work at different occupations like the French, after that we can little by little civilize them among themselves and with us."[25] Le Caron closed his letter by suggesting that Iroquoian peoples to the west who practiced agriculture offered the best chance for real conversion. Thanks to the writings of the lay brother Gabriel Sagard, it is the Récollet mission to these peoples, the Wendat, that is best known today.[26]

Little came of the Récollet effort to sedentarize the Innu of the Saint Lawrence Valley imagined by Le Caron, and their labors there seem to be chiefly remembered for an impoverished and half-finished seminary in Québec and their efforts to culturally and religiously convert indigenous children whom they sent to France.[27] Yet we can see in Le Caron's writing the constellation of horticultural ideas mobilized in the name of converting the Algonquian-speaking Innu communities of the lower Saint Lawrence. His writing emphasized the centrality of both restricting their spatial footprint and reimagining their ecological practices to the process of conversion. Le Caron echoed the language of Pierre Biard, who had similar ambitions for the Mi'kmaq at Port Royal and who had explained several years earlier that "when, little by little, their land is cultivated, they will derive from it their support."[28] By the time Le Caron penned this possible future for Innu peoples, other Récollets and the Jesuits to the east in Acadia had also come to see the adoption of agriculture as essential to the conversion of Algonquian peoples.[29]

Agriculture and horticulture provided a potent set of symbols to imagine the social worlds of Europe and to explore the relationship between questions of labor, class, and ecological practice.[30] Missionaries and colonists anxious to plant Christianity in New France looked to Native ecological practice as a material manifestation of their civility and readiness for religious conversion. Diverse aboriginal nations were frequently grouped into those who practiced agriculture and those who did not.[31] Marc Lescarbot's taxonomy is typical when he wrote that "living as the *Sauvages* do seems to me out of all reason. And to prove this, the following is an example of their way of living: From the first land (which is Newfoundland) to the country of the Armouchiquois, a distance of nearly three hundred leagues, the people are nomads, without agriculture, never stopping longer than five or six weeks in a place." He continued as he turned to the west, where "as to the Armouchiquois and Iroquois countries, there is a greater harvest to be gathered there by those who are inspired by religious zeal, because they are not so sparsely populated, and the people cultivate the soil, from which they derive some of the comforts of life."[32] Biard recapitulated this same distinction when he wrote, "As to the *Sauvages*, they know nothing about cultivating the land, and cannot give themselves up to it, showing themselves courageous and laborious only in hunting and fishing. However, the Armouchiquois and other more distant tribes plant wheat and beans, but they let the women do the work."[33] Ignoring or ignorant of the complex reality of ecological regimes that routinely blurred the line between the *sauvage* and the cultivated, these authors translated the mutually complementary ecological practices of Algonquian and Iroquoian peoples into a moral scale that indicated the ease with which they might be converted.[34]

The first writings of the Jesuits who traveled to New France in 1625 and again after the return of the colony to France in 1632 focused extensively on describing and unpacking the ecological practice of nomadic peoples.[35] Much of this concern was pragmatic. Paul Le Jeune had experienced the travails of a winter spent traveling with an Innu family and, even if one wonders from his description whether it was not harder on the family with whom he traveled, he recognized the impossibility of maintaining a "flying mission" among such peoples.[36] Le Jeune explained that "if I can draw any conclusion from the things I see, it seems to me that not much ought to be hoped for from the *Sauvages* as long as they are wanderers; you will instruct them today, tomorrow hunger snatches your hearers away, forcing them to go and seek their food in the rivers and woods. Last year I stammered out the Catechism to a goodly number of children; as soon as the ships departed, my birds flew away, some in one direction and some in another."[37] Le Jeune cast these "poor" people as subject to natural rhythms and seasonal cycles and as "idlers" unable to master either themselves or their environment.[38]

*   *   *

Understanding the Algonquian and Iroquoian cultures of the Northeast and the Great Lakes was no easy feat for early colonists and missionaries. Some of the Jesuits who came to French North America began their education in aboriginal cultures before they even left France, but there were limits to what could be learned an ocean away and in the absence of aboriginal teachers.[39] Most missionaries began an apprenticeship in Native languages and cultures when they arrived in New France. Copying manuscript dictionaries and working alongside missionaries who had been in the field for decades, newly arrived Jesuits also apprenticed themselves to aboriginal peoples who instructed them in their languages and cultures.[40] The study of these dictionaries and word lists reveals the daily proximity of Jesuit and aboriginal communities and suggests that the Jesuits' language instruction was part of a broader study of aboriginal culture.[41] Lalement reported receiving "entire narrations" from indigenous women of the Tionontate (Petun), a practice that was common as Jesuits looked to learn more than individual words.[42] As Jesuits sought to learn about aboriginal communities they watched and listened as much as they interrogated and became, in the process, deeply knowledgeable about many facets of indigenous culture.

Although it has become common to privilege the linguistic nature of "biocontact zones," French efforts to translate and appropriate indigenous ecological and botanical knowledge took place in a variety of forms.[43] Travel narratives often related knowledge gained through immersion in aboriginal cultures; this was knowledge acquired as it was lived. As often unwitting apprentices in aboriginal lifeways, however, missionary and colonial authors frequently found themselves subject to gender and social roles that determined their access to particular sorts of knowledge. As unmarried men and enemies of the many medicine societies and shamanic healers of the Algonquian and Iroquoian communities of the Great Lakes, for example, Jesuits were excluded from many of the existing circuits of knowledge transmission in these indigenous cultures.[44]

While missionaries were often left to observe the use of plants in indigenous lives and deduce their properties on their own, they also learned when their Native hosts explicitly taught them. Writing of his time among the Wendat in the 1620s, Gabriel Sagard recounted how both missionaries and lay French workers were taught the poisonous nature of two indigenous plants that he identified as *Ondachiera* and *Ooxrat*. Even here, however, it seems that these qualities were explained after the plants had been ingested and after nearby Wendat had been forced to intervene to save the lives of hapless Frenchmen. He recounted, for example, that "one day we had a great apprehension for a Frenchman, who after having eaten [*Ondachiera*], became greatly ill in an instant & pale as death, he was nonetheless cured by the vomitives that the *sauvages* made him swallow."[45] Where these aboriginal peoples of the Great Lakes demonstrated and explained the qualities of specific plants to French missionaries and colonists, it seems that the exchange was at least partially motivated by a fear that the French, if left unsupervised, were likely to hurt or poison themselves.

Accounts such as Sagard's suggest a recognition among missionaries of the limits of their own knowledge and the need to learn from those whom they otherwise presumed to teach.[46] Missionaries also demonstrated an increasing awareness of the cultural specificity of what they learned. Early in the seventeenth century, Sagard, author of the first Wendat dictionary, clearly believed that any number of words referred to a single salient object, writing that "for example, the Hurons call a dog *Gagnenon*, The Epicerinys *Arionce*, & the Canadians or Montagnets *Atimoy*: thusly can one see what a great difference there is between these three words, who signify nonetheless one single thing."[47] He was thus able to write some pages later that the word for "Grain of all sorts" was simply "Onneha."[48]

Even as Jesuits continued to rename both landscapes and converts, the more they knew of indigenous languages and cultures, the less confident they were in fully appropriating indigenous plants and knowledge. Not only would later dictionaries add additional information about the flora of French North America, they would prove that these sorts of facile translations were impossible for the plants used by indigenous peoples. For if the Jesuit Chaumonot was able, in one of his Wendat dictionaries, to continue to offer direct translations for trees such as apple, maple, and cedar, he nonetheless recognized over a dozen different types and preparations of corn that clearly stretched the limits of Jesuit botanical taxonomies.[49] Similarly hinting at the incommensurability of aboriginal and Jesuit knowledge, his colleague Jacques Bruyas identified a new type of wood found among the Mohawk as that "which is used to make torches to hunt turtles at night."[50] Working among the Abenaki in the eighteenth century, the Jesuit Sébastien Râles would include a "stinking wood to force vomiting" and, as further evidence of his inability to translate indigenous plants into European terms, would translate the fruit *Masiman* simply as "red, little."[51]

Missionary authors established a complex relationship with indigenous knowledge that was at once inquisitive and equally dismissive. Statements that evidenced clear-eyed appreciation of indigenous ecological knowledge followed quickly upon denunciations. Le Clercq, who described Mi'kmaq cultures where "all are by nature physicians," nonetheless spoke of a "repugnance" that had to be overcome to live among people who lacked even the sense of "ants and little squirrels, who by an instinct that is as equally admirable as it is natural, carefully amass necessities for the winter in the summer when they are abundant."[52] Gabriel Sagard wrote about "cheveux relevez" who "are errant, if not for several villages among them that sow corn," without ever question-

ing his larger categorization of "errant" peoples.[53] Perhaps it is best to say
that missionary authors gathered significant experience of indigenous ecological knowledge without ever coming to terms with the extent—or even
existence—of the indigenous knowledge systems that made living in American
environments possible.

In the absence of this broader theorization, missionaries learned indigenous
knowledges as they ate, as they were cured, and as they watched indigenous men
and women interact with their local environments. The knowledge of American flora expressed in their texts was indigenous botanical knowledge but
as it was observed in indigenous lives and as it had been distilled into discourse
for a European audience.[54] Sagard and later Jesuits laboriously described
methods of preparing corn for food among the Wendat, for example, while
others carefully noted the trees used to construct houses, canoes, or tools.[55]
As the product of an eclectic mix of observations and experiences, however,
Jesuit knowledge of American flora was therefore as likely to be unwieldy as
it was syncretic; it was rarely greater than the sum of its parts.[56] At the very
least, however, it seems a clear exaggeration to state, as Le Jeune did in 1634,
that missionaries believed that "the names . . . of an infinite number of flowers, trees, and fruits; of an infinite number of animals, of thousands and
thousands of contrivances, of a thousand beauties and riches, all these
things are never found either in the thoughts or upon the lips of the *Sauvages*."[57] Missionaries had expected to find *sauvage* places and people that
would justify conversion and colonialism, but they had instead encountered
knowledge that challenged their own and that stretched the limits of their ability to describe their new environments.

*   *   *

The Jesuit mission to New France aimed at a wholesale reorientation and re-
definition of human, spiritual, and ecological relationships that were understood to be mutually reinforcing in the creation and maintenance of Native
American environments. Jesuits in New France made explicit comparisons between the indigenous communities they served and those their confrères had
worked with in Paraguay.[58] In 1637, Le Jeune responded to a letter that asked
"for some enlightenment as to what we may hope for the establishment of the
Christian Religion, and then communication with the countries contiguous
to the *Sauvages*, their frontiers and boundaries." He responded that "if he
who wrote this letter has read the Relation of what is occurring in Paraquais

[Paraguay], he has seen that which shall some day be accomplished in new France."[59] He continued by emphasizing the comparison. "These peoples where we are," he wrote, "are exactly like those other Americans, called Paraquais."[60] In both cases, and indeed throughout the Jesuits' war on supposed irreligion the world over, authors such as Le Jeune championed the need to "reduce" non-Christians in a bid to civilize and convert them.[61]

A key facet of the conversion of the nomadic peoples of the Northeast and Saint Lawrence Valley thus focused on inducing them to become agriculturalists: a transformation of both ecological relationships and the gendered division of labor that saw agricultural roles as naturally feminine.[62] The indigenous people targeted for settlement at the first sites at Sillery and La Conception seem themselves to have understood the significance of agriculture to the missionaries and they responded with statements that suggested their willingness to take up agriculture.[63] In 1641, Le Jeune explained that the "captain" of an Algonquian community remembered that "we promised thee last year . . . that we would come and dwell a day's journey from your Settlement, as much in order to learn the way to heaven as to cultivate the land."[64] Nonetheless he was aware, as historians have been since, that it was at least as much anxiety about Iroquoian attack and commercial decline that drove these people to reconsider their relationship with the French, God, and North American ecosystems.[65] "I do not know . . . which we ought rather to wish them,— adversity, or prosperity; sickness, or health," he wrote. "For, if the healthy do not become wiser in time by one than by the other," Le Jeune continued, "some sick ones, at least during their maladies, give us in dying the assurance—or, at all events,—the hope of their happiness."[66]

At both of these sites, Sillery near Québec and La Conception near Trois-Rivières, Algonquian families were introduced to new social and ecological roles. Jesuits enforced a new sense of marriage, of parenthood, and of relationships between the genders.[67] Both settlement sites grew from places of seasonal occupation to homes for numerous Algonquian and Innu families. While La Conception's population peaked at 80 in 1641, Sillery grew throughout the same period and numbered as many as 167 inhabitants by 1645.[68] French laborers hired with the support of the reductions' benefactors cleared land and sowed fields to fulfill the ambitions of the missionaries who hoped that they "would work for the *Sauvages*, on condition that they would settle down, and themselves put their hands to the work, living in houses that would be built for their use."[69] Yet these laborers found much of their time taken with clearing forests rather than raising buildings and crops. For their part, the aboriginal peoples who had previously acquired cereals from the Wendat found agriculture beneath their dignity, and the silence of missionary authors on the success of their efforts to induce their charges to produce crops argues against their success.[70] The reductions showed little success cultivating an agriculture-based community of neophytes.[71]

The language of reduction evoked "conversion as a process of submission to the yoke of the faith."[72] Reduction spoke to a discourse of domestication and taming in a manner akin to the *sauvage* nature that missionary authors similarly targeted for change. When a young convert died at Québec in 1636, for example, Le Jeune explained that "his case makes us hope that there will not be found in these wildernesses a nature so ferocious that our Lord may not tame it by his grace when it shall please him."[73] When he wrote that "the time will come when they can be domesticated," however, Le Jeune was writing about local animals. He continued, as he explained to potential colonists about moose, that

> we shall make good use of them, having them drag over the snow the wood—and other things which we shall need; these Gentlemen are keeping three of these animals, two males and one female, and we shall see how they will succeed; if they become tame, it will be easy to provide for them, as they eat nothing but wood. In time, parks can be made, in which to keep Beavers; these would be treasure-houses, besides furnishing us with fresh meat at all times. For if one sees so many ewes, sheep, and lambs in France, although the Ewe generally bears but one lamb a year, I leave you to imagine how much more Beavers will multiply, since the female bears several.[74]

The aboriginal peoples settled at Sillery were themselves understood within a botanical discourse that made their connection to a wild and untended nature explicit. Paul Le Jeune explained in the *Relation* of 1640 that "when we first came into these countries, as we hoped for scarcely anything from the old trees, we employed all our forces in cultivating the young plants; but, as our Lord gave us the adults, we are turning the great outlay we made for the children to the succor of their fathers and mothers,—helping them to cultivate the land, and to locate in a fixed and permanent home; we still retain with us, however, some little abandoned orphans."[75] The subjugation of a "wild" place and a *sauvage* people was understood as mutually complementary domestication.[76]

It is conventional to speak of these efforts as failure. Success in converting the peoples of the Saint Lawrence Valley was directly related to decisions made in the 1640s, where Jesuit missionaries accepted and adapted to indigenous realities.[77] Focusing on the outcome of these experiments, however, risks missing the substantial knowledge acquired through the process of studying and engaging indigenous cultures—knowledge of indigenous ecological relationships and the local environments in which they lived that was quickly mobilized in print for European audiences eager to learn more about this New France on the other side of the Atlantic. If the goal of these missions was intercultural communication, we need not be surprised that settlements such as Sillery became sites of encounter between different systems of ecological knowledge.[78]

* * *

The push into the Great Lakes that followed the establishment of a Récollet mission to the Wendat in 1615 brought missionaries into contact with new environments and new ways of knowing them. Spared the harrowing experiences of those, like Le Jeune, who had wintered among Innu peoples in the harsh environments north of the Saint Lawrence, Récollets such as Gabriel Sagard and Jesuits such as Jean de Brébeuf encountered sizable villages with extensive areas under cultivation.[79] Brébeuf, for example, made the comparison explicit when he explained that "the cabins of this country are neither Louvres nor Palaces, nor anything like the buildings of our France, not even like the smallest cottages. They are, nevertheless, somewhat better and more commodious than the hovels of the Montagnais."[80] Missionaries to the Wendat also described fields of corn, beans, and squash as well as the use of seasonally available wild plants that enriched the diets of the Iroquoian and Algonquian peoples to whom they brought their faith and their charge to civilize and assimilate.[81]

Missions to the Algonquian people of the Saint Lawrence had sought to settle nomadic peoples, restricting their mobility and simplifying their ecological practice in the same movement. Those who settled among the agricultural Wendat and other peoples of the Great Lakes region faced a more complicated task: adapting to spaces that were clearly horticultural but in which their own practices had difficulty setting root. Sagard, for example, described the first garden of his order surrounded by "a little palisade" that was necessary to "remove the free access of *Sauvage* children, who seek only to do wrong for the most part."[82] This suggests that European agriculture in these early missions produced fortified islands in otherwise indigenous spaces, but this seems simply one part of a larger tension between adaptation to indigenous lifestyles and the isolation that some believed was necessary for real conversion.[83]

Missionaries likewise learned that much of the ecological knowledge of the Haudenosaunee and Wendat was feminine in nature and was deliberately kept from men, be they Native husbands and brothers or Jesuit missionaries. The complexity of aboriginal ecological practice was in effect hidden from the view of missionaries who either neglected or were not permitted to take part in the seasonal agricultural and foraging activities of indigenous women. It is well documented that agricultural crops were the material, intellectual, and spiritual preserve of indigenous women.[84] Even if they disapproved of what they saw as a dereliction of a duty central to the definition of masculinity, Jesuit authors noted that it was women who planted, maintained, and harvested the fields of corn, beans, and squash that were central to the economic and spiritual life of Haudenosaunee and Wendat cultures.[85] In the western reaches of the Great Lakes region, French fur traders were similarly forced to accommodate the desires of female indigenous consumers when they sought an undeniably feminine plant, wild rice.[86] Missionaries were aware that the knowledge to cultivate and collect these plants was restricted to aboriginal women. Gabriel Sagard informed his readers, for example, that "just as the little boys have their special training and teach one another to shoot with the bow as soon as they begin to walk, so also the little girls, whenever they begin to put one foot in front of the other, have a little stick put into their hands to train them early to pound corn."[87] Joseph-François Lafitau wrote that the cultivation of corn, beans, and squash was the work of "all the women of the village" and that it was overseen by a "mistress of the field" who distributed seed and coordinated planting.[88] In his *Moeurs des sauvages américains*, Lafitau described indigenous cultivated plants such as wild rice and sunflower under the general heading of "Occupations of Women."[89] In the nineteenth century, Lewis Henry Morgan wrote that the plants themselves were understood to be essentially feminine in nature and were known collectively as *De-o-há-ko*, or "Our life, Our supporters," underscoring the significance of these plants in Haudenosaunee life.[90]

Iroquoian and Algonquian women had a special relationship with both specific plants and the other-than-human beings who maintained and nurtured them. While Jesuits were careful to observe and record much of what they were able to see, it is clear they remained apart and excluded as men. Jesuits quickly appreciated that this was a world to which they could never fully belong and understood that women constituted both a unique source of information and a singular threat to the fate of their mission.[91] "At first we believed that it was only the young boys who were brought up in these stupid notions," wrote Claude Dablon from Sault Sainte Marie in 1670, "but we have since learned that the little girls also are made to fast for the same purpose; and we find no persons more attached to these silly customs, or more obstinate in clinging to this error, than the old women, who will not even lend an ear to our instructions."[92] Yet it was to these same women that Jesuits were frequently forced to turn when they sought ecological or agricultural knowledge.

The relative absence of information about ecological practices in Jesuit texts might suggest that the Society of Jesus simply was not as interested in ecological knowledge as it was in the political organization of indigenous cultures.[93] Indigenous studies scholars have argued, however, that European and Euro-American ignorance of the importance of indigenous women's ecological and cultural knowledge was an intentional product of aboriginal lifeways.[94] We can account for the relative absence of women's oral traditions and the underappreciation of female figures such as Sky Mother and her daughter Lynx in the writings of missionaries, travelers, fur traders, and contemporary anthropologists, for example, by recognizing that these stories were often too important to be told to men. The focus of European observers on the role of the male twins Sky Holder and Flint remained consistent not because it was the most important to Haudenosaunee cultures but because it was the most accessible to Jesuits as men.[95]

Jesuits were distanced from much of aboriginal ecological practice by their status as men in a world where women kept their knowledge to themselves. Claude Chauchetière, a Jesuit missionary who worked and lived at the mission at Kahnawake between 1677 and 1694, was also an amateur artist. Most famous for his image of the Mohawk convert Kateri Tekakwitha and his account of her life at Kahnawake, Chauchetière also wrote a *Narration* of the mission at Sault Saint Louis.[96] While the narration provided a rare glimpse into life at the mission over the course of almost two decades, one of the text's accompanying images hinted at the distance that separated Jesuit observers and female ecological knowledge. Titled "On travaille aux champs" (Figure 6), the image might seem to suggest that Jesuits enjoyed a daily proximity to aboriginal agriculture that would promote the exchange of knowledge about local plants. Yet while the visual details of the image demonstrated a familiarity

Figure 6. Claude Chauchetière, "On travaille aux champs," *Narration annuelle de la mission du sault*. AD Gironde, H 48. Courtesy of Archives départementales de la Gironde, Bordeaux, France.

with aboriginal ecological practice, it also suggested that Jesuits watched these practices rather than engaging in them. Though the image showed that Jesuits observed the actions of indigenous women closely (in this case showing how aboriginal women shaped the earth to plant corn and collected eggs from trees), it hinted at the limits that this mode of knowledge exchange placed on the transmission of medical, economic, or cultural usage. Positioned as a detached observer, Chauchetière illustrated the absence of discursive exchange and a lack of familiarity with the specific plants being used. Yet absent any of the problematic spiritual associations that could derail oral exchange, "On travaille aux champs" demonstrated why nonparticipant observation could nonetheless function more effectively than conversations with indigenous informants.

While Chauchetière was clearly able to personally observe aboriginal women at work in the fields, the social distance created by aboriginal gender roles often translated into spatial distance as well.[97] Indeed, while Jesuits and male colonists occasionally accompanied families on winter hunting trips, their participation in the harvesting expeditions that brought nutritional and culinary diversity to aboriginal cultures seems to have been scarce. Jesuits rarely saw and equally rarely described the ecological practices that took women beyond the sight of their villages. Drawing on seasonally available nuts, fruits, berries, and vegetables that paleoethnobotanical research has demonstrated constituted a major part of aboriginal diets in the Great Lakes year-round, aboriginal subsistence patterns were far more nuanced than the Jesuit record suggests.[98] Iroquoian and Algonquian peoples traveled fairly significant distances to collect the diverse assemblage of "wild" fruits and vegetables that have been recovered in paleoethnobotanical excavations.[99] Indeed, one examination of midden sites and storage pits associated with Wendat settlements argues that current ethnohistorical research based on missionary and colonial accounts that focused on the production of staple crops such as the three sisters and wild rice perpetuates an ignorance of the diversity of historical aboriginal diets.[100]

Missionaries were therefore best able to comment on the products of female agricultural practices when corn, beans, squash, wild rice, and assorted fruits and berries were brought back to villages and consumed, processed, or stored for later use. Early descriptions of the variety of methods of preparation and consumption such as those by Gabriel Sagard testified to a profound interest in the culinary traditions of aboriginal people. Knowledge about American flora was obtained as Sagard and others following him observed and ate with indigenous people. Lafitau, for instance, provided a description and analysis of the processing of corn at Kahnawake that stretched eight pages in a recent edition.[101] Another anonymous Jesuit author who wrote "a long article . . . on corn" revealed the depth of personal experience that individual missionaries could draw on as they sought to describe aboriginal botanical knowledge. Critiquing another Jesuit's account of aboriginal cuisine, his description included a discussion of the different types of corn and their preparation, and he wrote that "the ideas of roasted corn and of popcorn are not accurate. 1. it is not at all corn that is still in the ear and still green that one uses, but one uses it husked and quite ripe. 2. One does not at all grill it on coals but in hot ash, or in the sand that one has made redden in the fire, and that one packs down after, or even in a trowel or in an appropriate pot like that which we use to roast coffee with the difference here that one adds a bit of fat. 3. It is not at all a particular type that makes popcorn: of the totality of corn that one roasts, there is a part sometimes more sometimes less often that blossoms in the manner of a flower."[102] It was an incredible amount of detail that bore witness to years of experience. "Perhaps the subject does not merit the effort to write so much and with such a bad pen," this author acknowledged, "but in the end when you speak of things, you must describe them such as they are."[103]

Bressani similarly described Iroquoian houses that featured "great pieces of bark supported by beams, which serve to hold up their corn, to dry it in winter."[104] Jacques Marquette included information on how food was processed when he described the agricultural economy of the Illinois, an aboriginal confederacy that lived just to the south of the Great Lakes along the Mississippi River. He wrote that "their Squashes are not of the best; they dry them in the sun, to eat them during the winter and the spring."[105] The Jesuits who traveled with the Algonquian communities of the western and northern Great Lakes were allowed a unique perspective on aboriginal seasonal migrations and subsistence strategies that drew upon the ecological richness and diversity of the region. In 1674, the superior of the American missions, Claude Dablon, recounted that "our other missionaries among the Outaouais labor holily and usefully, each in his Mission. Within a year, they have baptized more than five hundred infidels; and, this summer, Father Bailloquet alone baptized in two months a hundred children and some adults, fully one-half of whom are sure of paradise. He gathered this harvest while the *Sauvages* with whom he was were gathering that of certain small blue fruits, on which they and the Father lived during those two months."[106] Dablon later increased the botanical detail and wrote that "Father Bailloquet also proceeds there, from time to time; but, as a rule, he lives with the Algonquin of lakes

Huron and Nipissing. He it is who, as I have related, lived for two months this summer, with more than a thousand *Sauvages*, on small fruits here called blueberries [*bluets*], which grow only on rocks or in rocky soil."[107] Bailloquet was able to comment on indigenous knowledge because he lived it. What was transmitted was a lived knowledge rather than an abstracted science.[108]

Out of the innumerable possible botanical discoveries to be made in French North America, Jesuits therefore focused on the objects that had been highlighted by aboriginal usage. Yet where botanical knowledge was not exchanged with indigenous peoples, Jesuits were often forced to be satisfied with access to the products that indigenous cultures produced from local flora. In a letter written by the Jesuit Chauchetière in 1694, the author made this clear as he explained to his brother that "I send you a piece of bread which has come from a place 500 leagues from here. It comes from the Illinois country; it is made of meddlers or services, and has a very good taste."[109] The botanical knowledge of the Illinois peoples was implied but remained out of reach in an account that, while highlighting the finished product and its natural source, remained silent on specific indigenous ecological or culinary practices that produced this bread.

While social and physical distance led Jesuit accounts of American environments to overemphasize the cultivation of indigenous staple crops, a conceptual gap in Jesuit discourse also left missionaries unable to see and explain the complexity of aboriginal ecological practice. In 1653, the Jesuit missionary Bressani betrayed the existence of this gap when he wrote that besides squash, the Wendat had "no other fruits but *sauvage* ones."[110] Jesuits failed to recognize what recent archaeological and anthropological work has clearly demonstrated: that both through the selective pressure exerted on plant populations by regular harvesting and collecting and occasional and localized burnings that created habitats for fruit and nut trees, many of the ecosystems of the Great Lakes in the seventeenth and eighteenth centuries were anthropogenic.[111] For example, whereas explorers such as Nicolas Perrot demonstrated an awareness that plants such as *folle avoine* were planted and managed, the Jesuit Claude Dablon described wild rice as "a kind of marsh rye which we call wild oats, which the prairies furnish them naturally."[112] The result was that while Jesuits recorded many of the agricultural practices of indigenous societies, they missed the multiple uses of "the mosaiclike landscape of agricultural fields, villages and ceremonial centers, managed woods, and larger expanses of forest that were used for hunting" that recent ethnobotanical research suggests were the norm throughout the seventeenth- and eighteenth-century Great Lakes region.[113]

Jesuits did in fact learn a great deal about aboriginal ecological knowledge and the use and character of North American flora, but it was a partial view of a complex knowledge system. If knowledge was learned as it was lived, there were definite limits to the reach of Jesuit experience. Jesuits were most often able to describe the product and practice of ecological knowledge in aboriginal villages where their missions were located, but indigenous cultures' ecological practices that took place out of Jesuit sight largely remained a mystery. Limitations on the transmission of ecological knowledge were determined by its gendered nature within the Algonquian and Iroquoian cultures of the Great Lakes. If it is nonetheless certain that knowledge of American flora circulated between indigenous and French communities in colonial North America, it is clear that the process of transmission was haphazard even if the results were nonetheless impressive.

*   *   *

As they settled among Algonquian and Iroquoian peoples from the Great Lakes to Acadia, Jesuits increasingly understood that these cultures traced the properties of plants to the intervention of what anthropologist A. Irving Hallowell has called other-than-human beings. Hallowell wrote of the Ojibwe with whom he worked that when considering these beings, "a natural-supernatural dichotomy has no place."[114] He further explained that in a culture in which "natural forces" as such did not exist, these beings were "neither the personification of a natural phenomenon nor an altogether animal-like or human-like being."[115] Instead, these were social relationships with beings who possessed the abilities (power termed manidoo by Algonquin peoples, orenda by Iroquoian peoples, and wašicun by Siouan peoples) to intervene and influence the human and non-human world.[116]

Jesuits frequently commented on efforts to propitiate the spirits of bears, beaver, deer, and natural sites such as lakes and waterfalls. Le Jeune's winter north of the Saint Lawrence in 1634 exposed him to the spiritual significance of their quotidian interactions with the non-human world. He noted, for example, that when eating "a young and very tender Beaver" one "should be most careful not to give the bones to the dogs, otherwise they believe they will take no more Beavers. They burn these bones very carefully. If a dog should eat them, there would be no more good hunting."[117] Le Jeune seemed to have understood that much of Innu life was concerned with making sure that intricate relationships with the natural world were maintained through proper ritual and social relations.[118]

Descriptions of the relationship with the powers involved in the growth and protection of wild and cultivated crops and the medicinal properties of plants were relatively rare.[119] Jesuits were nonetheless aware that there was a close affiliation between powerful plants and other-than-human beings, even if they misunderstood and thought that it was the plants themselves that were worshiped. Julien Binneteau claimed that the Illinois among whom he preached "claim that medicinal herbs are gods, from whom they have life, and that no others must be worshiped. Every day they sing songs in honour of their little manitous, as they call them. They inveigh against our religion and against the missionaries. 'Where is the God,' they say, 'of whom the black gowns tell us? What does he give us to induce us to hear them? Where are the feasts they give us?'"[120] In fact, the plants themselves were not worshiped in their own right. Aboriginal peoples instead were keenly aware that their encounter with a specific plant was mediated by their relationship with the other-than-human beings who were the ultimate source of the plant's power and the knowledge of how to properly use it.

At times in the Jesuit *Relations* it is not a matter of if these non-human forces are invoked that distinguished missionary from Native but what powers were invoked.[121] Missionaries, for example, suggested that practicing Catholicism was an integral facet of successful agriculture. Among the Wendat in 1642, Lalement wrote that "we do not know whether GOD willed to reward their Faith, and to punish the impiety of the others; but we were witnesses that most of the corn did not ripen, especially that belonging to those who had sacrificed to the Devil, while our Christians gathered a fair crop."[122] Only two years earlier he had encouraged his neophytes, upon "seeing there the fruits of the earth, uncommonly flourishing," to "kneel . . . and thank God for these good things which he gives us; it is the very least that we can do, since he continues his blessings upon us without ceasing."[123] The God that the Jesuits introduced to indigenous communities was an active one every bit as animate and involved in the world as the other-than-human beings that aboriginal cultures relied upon to maintain the workings of the natural world around them.

Iroquoian and Algonquian botanical knowledge was neither fixed nor universal, and the properties of the plants themselves were mutable and understood to be an effect of negotiations between the human and other-than-human worlds. Aboriginal peoples therefore understood that the exchange of knowledge with European colonists and missionaries necessitated educating the newcomers in how to maintain good relationships with the other-than-human beings that were responsible for the efficacy of medicinal plants and the growth and productivity of food crops. For particularly powerful plants, those most important for the well-being of the particular community, Iroquoian and Algonquian peoples renewed their bonds seasonally.[124] In ceremonies timed to coincide with seasonal markers such as the planting of agricultural crops, the arrival of green corn or strawberries, or the fall harvest, Iroquoian and Algonquian peoples offered thanks to the creator and other-than-human beings on whom they depended and renewed their bonds by offering these same powers specific objects that they desired such as tobacco.[125] It was not just the plants themselves that were cultivated; relationships with them were cultivated as well.

Iroquoian and Algonquian peoples of the Great Lakes understood that their own behavior influenced the botanical world.[126] An eighteenth-century account by Joseph-François Lafitau suggested that maintaining the appropriate respect for plants that could change their locations and properties lay at the heart of the relationship with the other-than-human world. Lafitau wrote that certain plants prized chastity and demanded it of those who would collect them for medicinal use. "They are persuaded that the love of this virtue extends as far as the natural sentiments of plants," he wrote, "so that there are [among them] those which have a feeling of modesty as if they were animate; and so that, to be effective in remedies or even when the diviners are not called upon, they expect to be employed and put to work by chaste hands, lest they lose efficacy. Several have said to me often, speaking of their illnesses, that they knew very well secrets for curing them but that, being married, they could no longer make use of them."[127] Where aboriginal peoples were willing to share their botanical knowledge with Jesuits, they understood that imparting knowledges of ceremonies and commitments to the other-than-human world was essential.

Relationships with the other-than-human world were maintained at an individual and societal level and included both those plants considered wild and others considered cultivated by Jesuit observers. Some knowledge was given to specific individuals in dreams or by close relatives and was contingent on keeping the details of the relationship secret from the uninitiated.[128] The account of the "Neh Gan-Da-Yah of the Fruits and Grains" that the

Moravian missionary William Martin Beauchamp recorded among the Haudenosaunee over a century ago detailed what these relationships could involve. These "Little People" that protected and nurtured plants started their annual work with the strawberry, where they "loosen the earth around each strawberry root, that its shoots may better push through to the light."[129] They made certain that the plants on which the spiritual and material survival of the Haudenosaunee depended grew and flourished. Throughout the growing seasons, they "are ever vigilant . . . and vigorous are their wars with the blights and diseases that threaten to infect and destroy the corn and the beans."[130] They were also, however, jealous of their privacy and agreed to continue their work only so long as the secret of their existence was not betrayed. If it was, it was feared that vines would freeze over the winter, birds would not travel south, and ground animals might forget to burrow.[131] Working among the Wyandot, descendants of the Wendat who were moved to Oklahoma in the nineteenth century, the anthropologist Marius Barbeau similarly learned that the efficacy of maple syrup harvesting depended upon indigenous women who kept a piece of magic sugar that had been given to them by the female spirit of the tree safe and to themselves.[132] If the relationship with specific medicinal plants was upset, it was expected that plants would hide from would-be collectors or, worse still, poison rather than cure the intended patient.[133] Plants that were common could quickly become rare. Those that remained might become slippery to evade an ill-intentioned or ill-prepared collector.[134] Plants demanded both respect and acknowledgment as they were collected and used.

Texts written by authors who had lived with indigenous peoples of the Saint Lawrence Valley and around the Great Lakes show that their hosts were wary of giving much of their botanical knowledge, whether economic, agricultural, or medical, to missionaries. Clearly establishing motives is difficult when this secret knowledge, by its very nature, escaped mention in many of the extant accounts of colonial encounter in French North America. Although colonial expansion and the antagonistic relationships between Jesuit missionaries and Algonquian and Iroquoian healers, medicine societies, and women limited access to indigenous ecological and botanical knowledge, it is overly simplistic to frame all limits placed on what Euro-American communities could learn as aboriginal resistance to colonialism. Indeed, nineteenth- and twentieth-century anthropological studies of Haudenosaunee, Wendat, and the Algonquian cultures of the Great Lakes suggest that secrecy was an integral aspect of indigenous cultures that obstructed the circulation of knowledge with colonial populations and within aboriginal communities themselves.[135]

In Iroquoian and Algonquian cultures, botanical knowledge was often only legitimate when it was secret, held by people who respected the larger ecological and non-human relationships in which they were embedded. The anthropologist William Fenton demonstrated that both men and women in Iroquoian cultures could possess herbal knowledge but recorded that his modern-day informants deliberately restricted the transmission of knowledge to members of their own family.[136] Knowledge was network specific and lineage dependent; Fenton found that even where the flora remained constant from one community to the next, usage varied according to the line of descent.[137] Fenton claimed that much of the knowledge that he was able to acquire from his Haudenosaunee informants came from tricking them into speaking about knowledge that they had intended to keep secret. "One of the surest ways to get an Indian's confidence is to get him talking about plants," he wrote. "He may freeze up if asked directly what he knows about the old medicines, but he is more likely to pluck some plant growing at hand, for example toad rush (*Juncus bufonius*), and relate that it grows everywhere along the village paths and that because it springs up when stepped on, the Senecas call it oge'o'dja'geon."[138] Even then, when the informant "Jim" collected a plant in Fenton's presence, he specifically asked the plant being collected to excuse the anthropologist's presence, stating, "Do not in the least restrain your power. And do not think that he is merely a white man."[139] Similarly, Wendy Geniusz suggests that among the Anishinaabe healers frequently prepared remedies in such a fashion that their ingredients could not be deduced by overly curious observers. Geniusz cites Huron Smith and Frances Densmore, both anthropologists who worked among Great Lakes cultures early in the twentieth century, who saw healers both grind medicinal ingredients to make them unrecognizable and add aromatic herbs to mask the scent of the remedy. Geniusz also claims that some Anishinaabe healers would not reveal the contents of a remedy even to their patients.[140]

Accounts of the transmission of this sort of oral knowledge were largely absent from travel narratives and mission relations. Where indigenous ecological knowledge such as this did surface in Jesuit accounts, it was often the result of the breakdown of extant circuits of knowledge transmission rather than their extension to include the French. When the Haudenosaunee woman Lute Andotraaon converted to Catholicism, for instance, the knowledge that had been imparted to her as part of her membership in a medicine society was made public as she renounced "a certain dance,—the most celebrated in the country, because it is believed the most powerful over the Demons to procure, by their means, the healing of certain diseases. Be this as it may, that dance is only for chosen people, who are admitted to it with ceremony, with great gifts, and after a declaration which they make to the grand masters of this Brotherhood, to keep secret the mysteries that are entrusted to them, as things holy and sacred."[141] Knowledge was legitimately transmitted when communities and families shared the stories of their origins and the history of their peoples, in the seasonal collection and farming that sustained indigenous communities, and through the ritual practice that renewed relationships with the other-than-human beings and the creator who were the ultimate source of the plants, their medicinal and nutritive properties, and the knowledge about their proper use. As the brief account above makes clear, the knowledge that was transmitted to Jesuits was embedded in ritual practice and the lifeways of indigenous cultures. Because the French were at the margins of much of indigenous ritual and social life, their access to the sorts of knowledge that circulated within medicine societies and through families was limited at best.

When this otherwise secret knowledge was gathered by Jesuit observers it was treated as cause for celebration. The Jesuit Louis Nicolas, who worked among the Odawa and other Algonquian peoples of the Great Lakes in the 1660s and 1670s, made his excitement at discovering a secret remedy plain to see in his *Histoire naturelle*. Describing the American oak, he stated that "the leaf of white oak is very good for healing wounds, and the Natives make their nails grow back with it after someone has torn them off with their teeth. I discovered this secret from a Virginian American whose nails were all torn out and who had one finger cut off and his arm stabbed. I would see him every day, going to the woods to cut white oak leaves."[142] He went on to write about not only where to collect the plant but how to prepare and administer it; this was a rare level of detail for missionaries often deprived of access to the medical knowledge of indigenous peoples.

Jacques Marquette similarly described, as part of a larger account of his voyage down the Mississippi, how he followed up on a clue left by an earlier missionary who had traveled to the southern Great Lakes:

> I also took time to look for a medicinal plant which a *sauvage*, who knows its secret, showed to Father Allouez with many Ceremonies.

Its root is employed to counteract snake-bites, God having been pleased to give this antidote against a poison which is very common in these countries. It is very pungent, and tastes like powder when crushed with the teeth; it must be masticated and placed upon the bite inflicted by the snake. The reptile has so great a horror of it that it even flees from a Person who has rubbed himself with it. The plant bears several stalks, a foot high, with rather long leaves; and a white flower, which greatly resembles the wallflower. I put some in my Canoe, in order to examine it at leisure while we continued to advance toward Maskoutens, where we arrived on The 7th of June.[143]

Jesuits pieced together indigenous knowledge not only from fragments that they had observed in practice but also from those they had read or heard from other missionaries.

In the eighteenth century, Pierre-François-Xavier de Charlevoix's frustration was evident as he discussed the tendency of aboriginal peoples to jealously guard their botanical knowledge. Describing a "fire dance" that he said he observed among the Mississauga in the journal of his expedition from New France to the Mississippi, he wrote that "I really wanted to know how a man could hold a lit coal in his mouth such a long time without burning it and without putting it out; but all that I could learn is that the *sauvages* are familiar with a plant which renders the part of the body that is rubbed with it insensible to the fire and that they have never wanted to give this knowledge to the Europeans. We know that garlic and onion could produce the same effect but for a very short time."[144] Given only fragments of hearsay, known or suspected botanical properties, and an observation of Mississauga practice, Charlevoix was left to reassemble indigenous botanical knowledge on his own. "One thing that constantly surprises me," he wrote, "is the impenetrable secrecy in which they keep their remedies."[145]

Where they were taught, it seems clear that indigenous informants frequently made missionaries aware of the larger spiritual contexts of the knowledge they held. As Jesuits discussed botanical knowledge with indigenous peoples with an increasing fluency, they understood that the exchange of botanical and ecological knowledge was fraught with danger as a result. Missionary accounts therefore suggest that the success of exchanges of botanical knowledge between missionaries and indigenous peoples could be inversely related to their competency in aboriginal languages.[146] Conversations with potential indigenous informants often complicated exchanges of botanical knowledge instead of facilitating them.[147] For example, in 1637 a Wendat shaman was willing to teach the Jesuits François Le Mercier and Jean de Brébeuf the medical use of "two roots."[148] The *Relation* of 1637 described the encounter in close detail where "on the 1st day of October, I felt some touches of illness; the fever seized me towards evening, and I had to give up, as well as the others." He recovered, but not before "Tonneraouanont, one of the famous Sorcerers of the country, having heard that we were sick, came to see us." The healer offered remedies, but "the Father satisfied him, or rather instructed him thereupon; he gave the sorcerer to understand that we could not approve this sort of remedy, that the prayer he offered availed nothing, and was only a compact with the devil, considering that he had no knowledge of, or belief in, the true God, to whom alone it is permitted to address vows and prayers."[149] As these Jesuits fully understood the spiritual contexts of Wendat medical knowledge, their discomfort grew. Rather than accept morally suspect knowledge from Wendat healers such as Tonneraouanont, Jesuits acted as intellectual gatekeepers, ensuring that their readers would remain in ignorance and refusing to profit from indigenous knowledge themselves.[150]

Jesuit authors saw something far more sinister at work in remedies and ecological practice that were often too effective to be entirely natural; while Jesuits represented bountiful crops and miraculous healings as evidence of divine favor for their mission, they were equally wary of possible manifestations of diabolical influence.[151] Missionaries such as the Récollet Le Clercq and the Jesuit Charlevoix were more worried when indigenous knowledge such as shamanic healing and predictions actually worked.[152] So while references to indigenous "superstition" were rife in the *Relations* and the later (and more skeptical) *Lettres édifiantes*, there was an underlying suspicion of aboriginal ecological and botanical knowledge that remained consistent throughout the seventeenth and eighteenth centuries. Bringing the standard of Christ to New France and representing their entire enterprise in the same militaristic discourse that framed missions in Europe, South America, and Asia, Jesuits were keenly aware of the human and supernatural enemies that faced them and searched the natural world and indigenous ecological knowledge for traces of diabolical influence.[153]

In the eighteenth century, the Jesuit Joseph-François Lafitau sought to distance his own work from the discourse of diabolism used by his confrères, but he nonetheless continued to show concern about the origins and moral status of indigenous environmental knowledge. Lafitau was as much concerned about the methods of indigenous knowledge systems as he was about their content. He wrote in his 1724 *Mœurs des sauvages américains* that "the basis of all the secrets of paganism has been that spirit of curiosity which leads men to wish to penetrate into the future or into the secret of the things which God, in his wisdom, has wished to hide in the secret places of his wisdom and the knowledge of which being above the natural forces, can only be made manifest to us by him, through the power of his goodness whenever he wished to give men some extraordinary mark of favor, or they can be imparted to us by the angels of darkness, by his divine permission and by virtue of the power He has given them."[154] This passage signaled an ambivalence in Lafitau's thought; he was not sure if indigenous botanical knowledge was simply the product of an illicit curiosity or if it was the product of a darker influence that his confrères had been chronicling for over a century. Regardless of its source, the natural knowledge of aboriginal cultures was not presumed to be innocent and was approached with considerable distrust.

* * *

In 1718, Lafitau lamented that when compared to the natural histories that had been produced in the Spanish Americas, his confrères had little to show for over a century of living with the indigenous cultures of French North America. In spite of the wealth of indigenous botanical knowledge described in missionary texts, Lafitau and historians since have been misled into believing that if knowledge was not spoken or written, it was not exchanged. Yet it seems clear that Jesuit missionaries learned a great deal about North American flora. They learned about seasonality and aboriginal life, about the proper methods for collecting and cooking the riches of indigenous fields, and about the extent of indigenous knowledge systems. Even as they resisted the temptations of botanical knowledge that seemed diabolical in origin, missionaries learned about the social lives of North American plants in indigenous communities and the ever-present other-than-human beings that gave them life and power. This was not knowledge, however, that often presented itself as such; it escaped the sort of anecdotal form that normally dominates discussions about the exchange of knowledge between Native and newcomer in the seventeenth- and eighteenth-century Great Lakes region and was rarely codified in the language of botany or natural history.

Interwoven with descriptions of the daily experience of living in aboriginal communities, accounts of the botanical practices of indigenous peoples were conveyed in multiple genres and to multiple audiences. As statements about the gendered division of labor in indigenous societies, as finely grained analyses of aboriginal cuisine, or as blow-by-blow accounts of their spiritual battles with shamanic healers, Jesuits communicated indigenous botanical knowledge to Europe as fragments of aboriginal lifeways. Jesuits nonetheless found themselves ill-equipped to translate the spiritually suffused botanical knowledge of aboriginal cultures and, as men, were excluded from many of the circuits of indigenous knowledge systems.

Missionaries acquired this knowledge because of their calling to cultivate a New France in North America. As both metaphor and ideology, cultivation encouraged an active engagement with both aboriginal peoples and their environments, and it promoted an empirical investigation of the root of American difference. Cultivation undeniably sought to efface this difference, but missionaries soon encountered opportunities to experience new cultures and to begin to appreciate the complexity of indigenous ecological knowledge. Jesuits and other missionaries never stopped seeking to domesticate indigenous peoples; nor did they stop seeking to undermine indigenous religious practice. Yet by the end of the seventeenth century, they had become aware of the limitations of a worldview that split the world into civilized and *sauvage*, and they participated in a broader effort to reimagine the relationship between the Old World and the New. Cultivation, an expansive and assimilationist legitimation of French colonialism, instead provided an education in the limits of European knowledge and ecological practice.

CHAPTER 4

The Limits of Cultivation

On November 7, 1712, the surveyor Gédéon de Catalogne presented a *mémoire* that synthesized the work he had done surveying the Saint Lawrence Valley in the preceding five years. The author's ambition was nothing less than that its reader "know Canada better than those who have frequented it for several years."[1] Catalogne's careful hand detailed the footprint of French colonial expansion along the river, describing where the forest had been pushed back, fields brought under cultivation, and homes erected. Yet it also reflected on spaces where colonialism had advanced only partially or had in fact retreated in the face of social and environmental challenges. If Catalogne's survey manifested the ardent desire to know and understand New France a century after the first establishment of colonies in the Saint Lawrence Valley and Acadia, it therefore also revealed the significant epistemological challenges that such a project faced. Even as Canada became clearer as a geographical space, it was in the context of a broader uncertainty about the nature of New France.

This was an era when a number of efforts were taken by the French state to make New France a legible and manageable space.[2] Catalogne's efforts to survey—to place—New France were part of a broader effort to come to terms with the experience of a century of French colonialism in northeastern North America and to impose order upon an increasingly indeterminate colonial map. His particular focus on Canada, an emergent cultural and legal space used to define those areas of New France that were most recognizably French, locates him as part of a wider effort to delineate French colonial spaces from those increasingly identified with indigenous peoples and unruly woods. Catalogne had arrived in the colony in 1683, a Protestant soldier who quickly took part in the 1684 attacks on the Haudenosaunee and worked his way up the colonial ranks through the wars of the League of Augsburg (1688–97) and of

Spanish Succession (1701–14). His success and skill as a surveyor attracted the support of intendants Jacques and Antoine-Denis Raudot.[3] The surveyor established himself in the midst of a cartographic revolution in the colony, when skilled geographers and engineers such as Guillaume Delisle and Robert de Villeneuve refined the representation of the colony for American and metropolitan audiences. New maps such as Jean-Baptiste Franquelin's *Carte des grands lacs* (1678) incorporated knowledge acquired from western exploration, the expansion of the fur trade, and the evangelization of the indigenous peoples of the Great Lakes watershed.[4]

These new techniques of precise calculation and representation made possible a dramatic reconceptualization of colonial space.[5] Where at the beginning of the seventeenth century, explorers and settlers such as Samuel de Champlain had pushed at the margins of their own maps and imagined empire on a continental scale, by the end of the eighteenth century, new maps testified instead to a retrenchment of colonial space and considerable uncertainty about the legibility of those places and peoples that fell outside of a colonial *oikumene* increasingly identified as Canada. It is also important, however, not to overstate the precision with which lines between the cultivated and the *sauvage* were drawn in this period. Considerable debate about climates, ecosystems, and the potential of American environments agreed only that ecological knowledge was essential to understanding the prospect and purpose of French colonialism in North America. Catalogne's survey seemed to present an image of manageable, legible space, yet his written accounts of North American environments demonstrated a fundamental anxiety about how to know New France a century after its founding.

The debate about where to locate Canada was therefore a more fundamental dispute over the limits of cultivation: the geographical limits of French colonialism in a continent far larger and far harsher than initially anticipated and the limits of cultivation as a means through which to reveal the true nature of *sauvage* landscapes. If colonial space became easier to plot, it nonetheless became much harder to know. Epistemological anxiety figured prominently in administrative correspondence and *mémoires*, in travel and missionary literature, and throughout personal communication that crossed back and forth across the North Atlantic.[6] Following debates about where to locate Canada can also allow us a unique perspective on how French colonial political ecology had changed after nearly a century of encounter and experience. The ideology of stewardship that had invoked the orderly world of an agricultural estate to represent the peaceful cooperation of natural and human agencies under the aegis of a benevolent patriarch had relied upon an expectation that any resistance to the imposition of French ecological regimes would be short-lived; climates would moderate and American flora and fauna would flourish as features of domesticated landscapes that brought forth their true potential. This georgic vision had been undermined by a century of cold weather and an inability to successfully introduce recognizably French ecological regimes. The naturalness of French colonialism itself—its ability to bring the latent, natural qualities of American environments and peoples to their fruition—was reconsidered and refigured throughout this period. Catalogne's ambition was therefore also an exercise that aimed to identify the conditions in which French colonialism could take root in eighteenth-century North America.

*   *   *

The work of Catalogne and others, like him, who sought to make France's North American colonies more legible was a measured response to the challenge of knowing a continent that had become less, rather than more, familiar over the seventeenth century. The delineation of legal and social spaces proceeded apace with a growing recognition that the expansive vision of earlier colonists who had imagined that American difference might be effaced could no longer be supported. Indeed, the landscapes described by Catalogne denied the easy intellectual assimilation of Champlain or Boucher and worked in tension with the neat legal geographies that his maps visualized. The dwelling perspective that had previously highlighted affinities between the flora of New France and Old now qualified French knowledge and localized colonial experience (Figure 7).

In the accounts of an author such as Champlain, colonial experiments with local flora had provided augurs for the conversion of indigenous places and peoples. In Catalogne's account, however, the directionality of exchange seems harder to trace, and it is the porosity of Laurentian cultural and ecological landscapes that he most effectively highlighted. If he continued to attest to the importance of familiar types of flora such as pine, oak, maple, and other useful trees, for example, Catalogne undermined the work that these generic names did in translating European knowledge into a North American context. About spruce, for example, he explained that "there are some that at their highest extremities grow a type of mushrooms . . . that the *habitans* call *guarigue*, while among the *sauvages* it is used for illnesses of the chest and

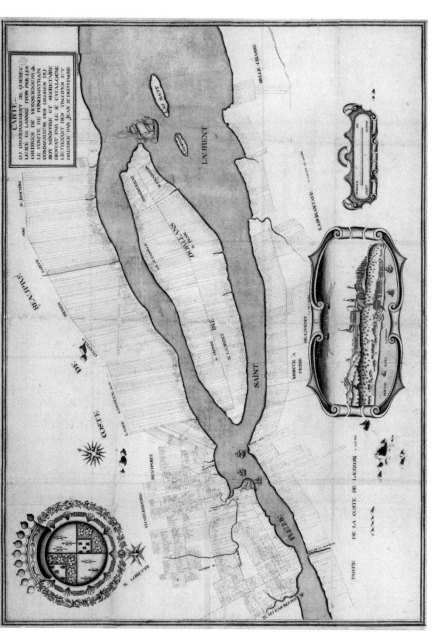

Figure 7. Gédéon de Catalogne, "Carte du gouvernement de Québec," 1709. Département Cartes et plans, GE SH 18 PF 127 DIV 2 P 2. Courtesy of Bibliothèque nationale de France.

for dysentery."[7] Familiar trees could become new in American ecosystems. He similarly described fields in which "almost all the *Sauvages* and even the French sow a type of pumpkin much smaller than that of Europe.[8] "French melons" were also grown by indigenous peoples.[9] These accounts therefore resisted the easy dichotomy of *sauvage* and cultivated that had rooted the political ecology of seventeenth-century New France, giving way, instead, to a more complicated acknowledgment of the limits of French knowledge and the realities of indigenous practice.

This might seem to echo the idealized hybridity envisioned through discourses of cultivation where French and indigenous flora coexisted under the benevolence of a colonial patriarch.[10] Yet even as Catalogne described mixed landscapes, he confidently pointed to firm differences between American and French flora. He was one of several authors at the end of the seventeenth century to highlight the successful introduction of species of plants that had originally been identified as growing *sauvage* in the colony. Louis Nicolas, for example, described the "seeds that are found in, or that have been brought into the Indies, and which produce abundantly there" and included "millet, oats, barley, rye, lentils, chick peas, beans of all kinds, peas of all sorts, [and] wheat of all kinds," among others.[11] The Baron de Lahontan, another late seventeenth-century traveler to the Great Lakes region, included a similar list that also juxtaposed familiar and foreign plants in such a manner as to simultaneously suggest both proximity and frequent differences. The page in this printing was centered on those plants that were "like in Europe," foregrounding the similarity of the plants visually far more than delving into their unique identities. Yet this identification subtly emphasized that much of North American environments was not "like in Europe."[12] Catalogne, for his part, listed the fruit trees and grains "from Europe" after going into greater detail about the trees, plants, animals, and fish that could be found in the region that he had been sent to survey. Among them were a number of the fruits that explorers and early missionary authors had hoped to cultivate locally, including cherries, prunes, apples, and pears. Perhaps most telling, however, was his brief listing of "vine" as one of the successful transplants in the colony.[13] While Catalogne still made mention of "vine *sauvage*" growing, for example, in "an infinity," he joined others in noting the successful transplantation of French vine instead of the cultivation of local varieties.[14]

Wild vines had, for explorers and missionaries, offered both a powerful symbol of French colonialism's aspirations and a source for wine: a necessity of French colonial lives.[15] Wild grapes were likely the most frequently referenced *sauvage* plant that early authors had highlighted as evidence for the need of French colonialism.[16] Yet as they realized the limits of their knowledge of North American flora, French colonists also began to appreciate their limited ability to impose European ecological relationships and transform North American plants by incorporating them into ordered gardens and fields. If early authors such as Champlain, Boucher, and Jesuit missionaries were optimistic about the ability to transform *sauvage* grapes into recognizable plants with palatable wines, eighteenth-century authors increasingly recognized both the diversity of grape populations throughout North America and the failure of efforts to cultivate them in the manner of their French counterparts.

By the end of the seventeenth century, colonists had largely given up trying to cultivate grapes that, Louis Nicolas explained, were capable of "producing only black grapes from which one can make wine, but so coarse that it thickens like mustard." Wine could be made, certainly, but it was "impossible to drink . . . without adding three or four quarts of water to a quart of wine."[17] The inferiority of American grapes could be diluted but not overcome. In the eighteenth century, Pierre-François-Xavier de Charlevoix described the wholesale removal of these local grapes from landscapes where they were now seen as impediments to settlement. When he wrote about Île d'Orléans, he explained that "when Jacques Cartier discovered this island, he found it full of vines and named it the Island of Bacchus. This navigator was Breton; after him came the Normands, who pulled out the vines, and substituted Pomona and Ceres for Bacchus. In effect it produces some good wheat and some excellent fruits. We have begun to cultivate tobacco there as well, and it is not bad."[18] This was confirmed by the Jesuit Jean-Pierre Aulneau, who wrote in 1734 that "the Île d'Orléans, the environs of Québec and the coasts a hundred leagues beyond are very well cultivated, and with the exception of vine, one finds there all that is in France."[19]

What was true at Québec is evident throughout the Saint Lawrence Valley, New France, and North America more generally. In Montréal in 1694, the Jesuit Claude Chauchetière wrote to his brother that "wine will be made this year" because "close by is a vineyard . . . which yields French grapes."[20] Landscapes in and around Montréal—a region that offered a growing season as much as a month longer than that of Québec—soon became home to vineyards and orchards of introduced fruit trees maintained by individuals as well as by organizations such as the Hospitalières.[21] This suggests that transplanting and the practices such as grafting that were essential to the establishment of introduced fruit became more important than efforts to cultivate *sauvage* flora.[22] That this was visible in some of the most prominent gardens of the colony suggests a broader transition that is otherwise difficult to trace in the archives.

For the travelers and colonists who pushed west in the eighteenth century, there was little sense that American environments presented familiar flora waiting for French cultivation. When in 1749 Gaspard-Joseph Chaussegros de Léry traveled to Detroit from Montréal, he "prepared kernels, seeds and grains of all species to furnish" the colony. He took with him

> French vines that he put in an iron barrel . . . [and] that he main-
> tained until Detroit where he planted it in the king's garden where
> they multiplied in such a fashion that several habitants have some
> in their kitchen gardens. There is reason to believe that if this
> plantation was encouraged, at Detroit we could make wine as
> good as that that one finds in several provinces of France, the
> softness and the beauty of the climate should engage us to culti-
> vate this plant for then, this will be the most productive country
> of America. . . . In the end, all that is lacking in this beautiful
> country are people who will be more concerned with its settlement
> than with commerce.[23]

This was still, clearly, an ecological vision of settlement but one that under-stood that colonists would bring both new ecological practices and the sorts of things with which they had built their old lives. Successful settlement would be a full transplantation of Europe rather than a cultivated America.

In these newly settled and newly explored regions, it became possible for grapes to become more indigenous as they became less French. The anony-mous author of the "Relation de la Louisianne," for example, outlined the lengths to which French colonists had gone to domesticate native grapes along the Mississippi and showed an awareness of indigenous classifications of the plant there. He recognized that "there are three types of them," each with its own taste profile, ecology, and potential for cultivation.[24] He explained that "there is one that the *sauvages* call *succo*, which has a leaf like the sycamore and has a similar fruit, like little prunes without bunches: the skin is very thick and the inside of it is not exquisite. The other is called *panko* and the leaf is like the *chasselas*, it has bunches like those in Europe . . . and always an acrid taste. The other carries the name *Panco*, the *sauvages* do not recognize any difference, but there is one nonetheless."[25]

The author continued by explaining that while local indigenous peoples made loaves of the fruit, French efforts to produce palatable wine had been frustrated. The text did not argue that such efforts were in vain, only that they would take a great deal of time and effort. He wrote:

> I saw some Frenchmen who wanted to make wine from them. They were obligated to add water, the liquor which came out of the grape was naturally very thick and could not flow, and this wine is good only lightly fortified because as soon as it is fermented it turns bitter. Maybe if it was well cultivated it would be better after it had been pruned. Though I knew a captain named Mr. de la Tour Vitrac, a gentleman from Brive, who has tried to prune them at his *habitation* on the coast at mobile, he even tried it for several years without success. Mr Diron commandant of the same post did the same thing as unsuccessfully at his habitation. . . . what he did wrong was to transplant it. . . . My opinion is that this vine growing naturally in the woods, covered in leaves for centuries, sheltered there from the winds, from great heat and from the injury of time; . . . being transplanted to a *habitation* which is exposed to the open air . . . changes its natural being . . . , one shouldn't be surprised if it no longer yields like it did beforehand. It can only be after many years, accustoming itself to this type of climate change that it could become capable of bearing good fruit, and being cultivated by old men who would have studied the climate and properties of each season.[26]

Successful cultivation required the adaptation of both colonists and colonial environments; it was only "old men" who have themselves presumably become seasoned who would be able to coax these grapes into a Frenchness that seems less and less inherent and less and less natural.[27] Otherwise, the sole success that this author recorded was the growth of Muscat grapes that had been brought from France and transplanted.[28] Like the missionaries who found that tending the figurative vineyard of the Lord by converting aboriginal peoples to Christianity had been a far slower process than originally expected, those who worked with the *sauvages* grapes of French North America found their initial ambitions upset or at least postponed.

Over the seventeenth and eighteenth centuries, even plants that had initially seemed familiar became more American and more closely associated with indigenous cultures.[29] The history of the American persimmon, for example, demonstrates how a plant's novelty could be discovered only gradually in these western spaces.[30] Marc Lescarbot was the first French author to describe the plant in what became French North America. His description was spare, and he wrote only that, in Acadia, "there is a kind of medlars, the fruit of which is bigger and better than that of France."[31] Yet by the time the French began to explore the Mississippi River watershed and discover this plant in greater numbers, they recognized that this French name did not quite hold. Henri Joutel displayed unease with an overconfident identification of the plant when he described "a sort of Fruit they call *Piaguimina*, not unlike our Medlars, but much better and more delicious."[32] Also in the Illinois region, the Jesuit Jacques Gravier wrote in 1701 that local indigenous peoples brought him "a large platter of ripe fruit of *piakimina*. It is pretty much like the French medlar."[33] If the identity of a plant was at least in part determined by what could be done with it, French authors embedded discussions of aboriginal ecological practice in their discussions of newly discovered plants.

If we return again to Catalogne's description of Canada, we can begin to see how aboriginal voices and indigenous ecological knowledge became an avenue for new understandings of once-familiar plants. In his *mémoire* on Canadian flora, he was open about the limits of his knowledge and where he recognized that indigenous knowledge could offer additional perspectives. "I will not describe," he explained, "an infinite number of plants and simples of which the properties are almost only known by *Sauvages*."[34] Catalogne had abandoned any pretense that American plants were simply wild versions of familiar European species. He described, for example, "Pemina," which "is a shrub that grows along the streams and prairies that produces vibrant red fruit in bunches." He also noted the presence of "Bluëst" and "Latoca" (blueberry and cranberry, respectively), which were among the first indigenous plants described by European explorers to northern North America.[35]

Exploration and contact with indigenous cultures encouraged French authors to reconsider whether they were in fact seeing familiar flora or, as they would ultimately suggest with medlars, new species that only happened to look something like French types.[36] Recovering the histories of these borrowed words runs up against the same challenges posed by any study of borderland regions where cultural contact and exchange were undoubtedly frequent but are, today, poorly documented in written records. Nonetheless, there are real glimpses of the inclusion of indigenous names for plants among authors whose texts foregrounded knowledge produced through daily experience of novel species. The lexicography of the Jesuit Pierre-Philippe Potier, for example, contains numerous indications of this sort of transfer toward the end of the French regime.[37] The scholar Marthe Faribault, for example, counts ten distinct references to indigenous-language plant names in the missionary's "façons de parler." These include *atoca* "that is found under the snow," a "cornar" that was a "seed that attaches to clothes," and "pacane" (pecans) that were also described by a number of contemporary authors.[38] Potier included these terms alongside more familiar French names and seemed to demonstrate, again, a transitional moment in European encounters with American flora. His "figue," for example, was ill-suited to the name, which he acknowledged by defining the fruit as "enveloped in a little sac, of the form of spider web," that the missionary "ate at the portage before arriving at onanguissé bay."[39]

This was only part of a broader incorporation of indigenous language terms to describe the natural and social worlds at the margins of French settlement.[40] Nicolas Perrot described a plant that he encountered in the seventeenth-century Great Lakes when he wrote that "the *Sauvages* call this root *Pokekoretch* in their language, and the French do not give it any other name because it is not seen in Europe."[41] Where plants were demonstrably new, the Illinois colonist Pierre Deliette claimed that aboriginal language names could even be preferable to French neologisms with limited appeal to wider audiences. Deliette wrote, for example, that in the Illinois region "there were other trees as thick as one's leg, which bend under a yellowish fruit of the shape and size of a medium-sized cucumber, which the *sauvages* call *assemina* (Pawpaw Tree). The French have given it an impertinent name."[42]

Experience with indigenous knowledge made French colonists and missionaries reconsider what they knew about American plants. Beyond providing the French with a new source of sweetener and a potential commodity, for example, learning about maple trees from indigenous communities encouraged French authors—travelers, colonists, and missionaries alike—to reconsider how they studied and used the tree. Missionary authors—the Récollet Gabriel Sagard and the Jesuit Paul Le Jeune—first described the indigenous practice of gathering sweet sap in the 1630s.[43] By the time Pierre Boucher wrote his *Histoire veritable et naturelle* in 1663, he identified the maple as the tree that produced a "quantity of water, which is sweeter than water with sugar mixed in."[44] Yet this soon provoked the question: Why, when maples were noticed across the continent, did they produce syrup and sugar only in certain areas? Could this mean that, given the right conditions, European maples were also capable of producing syrup or sugar? Certainly, some authors assured that, even if less productive, other trees could also produce syrup.[45] Louis-Armand de Lom d'Arce de Lahontan, who traveled in the region at the end of the seventeenth century, was one of the first to argue that the trees that were called maples in North America were in fact different from those of Europe. "The maples are almost the same height and width," he explained, "with the difference that their bark is brown and the wood reddish. It has no rapport with that of Europe. Those of which I speak have an admirable sap."[46] The amateur botanist Pierre-Joseph Buc'hoz wrote that experiments with European maples had proved that they could also produce syrup but that no European varieties were as productive as American maples.[47]

Yet considerable uncertainty remained. Charlevoix pointed to ecological differences between North America and Europe and suggested that "our maples might have the same virtue, if we had as much snow in France as in Canada, and if it lasted as long."[48] J. C. B., a soldier and author of a diary describing the eighteenth-century Ohio valley, informed his readers that it was important to ensure that it was cold the night before the sap was collected and to pierce only the side of the tree that faced the sun.[49] Even botanists such as the royal physician Michel Sarrazin had trouble explaining why some maples produced so much more syrup than others. Yet Sarrazin, who ultimately sent specimens of four different newly identified species of maple to the Jardin du Roi in Paris, was able to identify a particular species (which he called *Acer Platanoides*) and climatological factors that explained why "there are maples whose sap produces no sugar at all. There are maples that do not produce much. Finally there are some that do not produce it predictably."[50] Sarrazin's contention that it was both biological and climatological difference that separated European and American maples points to considerable ambiguity about the roots of American difference. For Sarrazin and others, a new botanical language of species and genera became crucial for representing an essential difference between French and American flora that could not simply be cultivated out or even replicated abroad.

No longer were American plants assumed to be *sauvage* versions of those that existed in France. While the botanical science then emerging in Europe could provide powerful tools for describing (if not explaining) floral difference, it was not so easy for colonial authors who struggled to explain their surroundings. Familiar names frequently proved insufficient, and the experience of American differences forced reconsideration of the transatlantic relationships that these names implied. Experience opened up a gap between Old World templates and New World realities. New France now threatened to become more new than it would be French.

*   *   *

The emergence of Canada as a distinctive cultural and ecological space at the end of the seventeenth century must be understood within the context of this epistemological anxiety. If Catalogne's objective was to enable his readers to "better know Canada," his task was made easier by the fact that his was a much smaller Canada than had been described in the preceding century. It was a colony increasingly defined by the legal geography of seigneurialism.[51] The engineer's attentive study of territorial boundaries provides one of the clearest views of colonial settlement along the Saint Lawrence and gestured toward a metropolitan vision of the colony where seigneurial boundaries demarcated human space from an uninhabited wilderness. It was a survey that nonetheless populated French landscapes with indigenous plants and peoples and that therefore also emphasized the ecological and cultural porosity of these same legal boundaries.

Catalogne's methods for representing French North America differed substantially from those of seventeenth-century surveyors.[52] The texts and maps of early missionaries and explorers had represented the same region as a beachhead from which to reclaim a *sauvage* continent; they made the argument that a New France could be cultivated in North America with the introduction of French social and ecological relationships. By the eighteenth century, few authors made claims that cutting forests would transform the climate any longer or that cultivating *sauvage* plants would make them French.[53] The geographically expansive vision that had first encouraged French settlement in North America was cast into doubt as administrators, explorers, merchants, and missionaries debated instead where boundaries might effectively be drawn to allow for the preservation of more definitively French spaces. Just as faith in cultivation had encouraged missionaries to venture into indigenous communities that opened them up to new ways of inhabiting and creating American environments, the experience of trying to cultivate a New France had shown colonists and colonial officials the limitations of their ability to produce French places in North America.

Canada emerged from Catalogne's *mémoire* as an assemblage of entangled histories that told the story of these limits; each seigneury was the product of a specific configuration of human and natural forces. Yet any pretense toward a coherent and self-contained colony was undermined by Catalogne's description of seigneuries that had been "abandoned" or that had insufficient soils or labor to realize the promise that their boundaries seemed to offer.[54] The seigneury of St. Sulpice had soils that varied substantially in their quality and had not been well established because of the threat posed by Haudenosaunee attack.[55] The seigneury of Saint Ours suffered from mediocre soils and "negligent" habitants.[56] The study of place was then necessarily an interdisciplinary pursuit, focused on understanding the complex interplay between the social, political, economic, and ecological realities of the colony.

Figure 8. Robert de Villeneuve, *Carte des Environs de Quebec en La Nouvelle France*, 1688. Département Cartes et plans, GE SH 18 PF 127 DIV 7 P 5 D. Courtesy of Bibliothèque nationale de France.

The surveyor's *mémoire* echoed other studies of the Laurentian valley such as the *Carte des environs de Quebec en La Nouvelle France* (1688) by the royal engineer Robert de Villeneuve (Figure 8). In an elegant green (which improved upon a previous 1686 version where forests were marked in brown), de Villeneuve's map "very exactly" traced out the uneven footprint of settlement near the colonial capital. To our eyes, it confirms the insights of historical geographers who have emphasized the importance of waterways and good soils, as well as the undeniable agricultural growth that was accomplished through a gradual expansion from existing settlements. It also represents the unevenness of the process of cultivation as, near the bottom of the map, several "abandoned clearings [*déserts*]" are immediately visible.[57] Taken together, these documents speak to an awareness of the precarity of the project to cultivate a New France in North America in the early eighteenth century.

Catalogne was not the first to use the name "Canada" to describe parts of New France, nor was he the first to use the term "Canadian" as an adjective for colonists born of French parents (that honor belongs to religious authors).[58] He was one voice among many in the years in which colonial societies were emerging in French North America.[59] Catalogne's survey is nonetheless particularly useful for foregrounding the influence of colonial environmental history on these conversations and on emphasizing their essentially spatial and ecological character. His survey was, in effect, a careful assessment of the footprint of French colonial expansion along the Saint Lawrence River. The studious attention to the boundaries of each plot was obvious in accompanying maps that visualized the order projected onto Canada's ecosystems.

It was only at the end of the seventeenth century that colonial-born French speakers became known as Canadians, and this was a period in which the nature and practices of Frenchness in North America were seriously considered. The remapping of colonial space was often effected at the level of individual bodies and families. The language one spoke, the foods one ate, the clothes one wore, and the spouse one took became visible markers of shifting cultural and social boundaries in French North America.[60] If it is now less common to gesture toward an eighteenth-century golden age that saw the first glimmers of French Canadian national awareness, this was nonetheless a period in which new ways of understanding and experiencing the American continent were being articulated.[61]

Ambiguity about where Canada actually was—and, by consequence, about its ecological character—had reigned since Jacques Cartier had first landed in the sixteenth century. In his documentation of his voyages, Cartier named the region at the mouth of the Saint Lawrence Canada and the new people with whom he interacted "Canadiens."[62] He thus described a stretch of the region between Trois-Rivières and Québec to the west and Grosse Isle to the east.[63] Throughout the sixteenth and early seventeenth centuries, the term remained essentially fluid. Differences in the pluralization of the toponym, uncertainty about whether the term described a river, a province, or a country, and the term's uncertain relationship with the Iroquoian peoples of the Saint Lawrence region who had disappeared between the time Cartier and Champlain explored the area ensured that the location of Canada remained unmoored to a specific colonial or indigenous space.[64] Early maps were equally uncertain where to place the name, with the geographic focus of Cartier quickly expanding to include far larger regions of the continent than the explorer had initially named.[65]

The location of Canada—and with it its counterpart New France—could remain simply unspecified. These were less precise geographical markers than a generic reference to parts (or all) of French colonial North America.[66] In the commissions provided to would-be colonizers in the court of Henri IV and Louis XIII, Canada was frequently listed as one of several *pays*, or lands, in France's New World.[67] This was in spite of the fact that Champlain's 1613 map located Canada to the south of Labrador and to the east of the mouth of the Saint Lawrence at the same time.[68] Even if authors knew where Canada began, they could also be hesitant to locate where it ended. While Gabriel Sagard's *Histoire du Canada* (1636) included accounts of the Récollet missions to the Wendat in what would become the *pays d'en haut*, he located them all in a larger Canada. He included reference to Canadians as a distinctive linguistic group that we would now call Algonquian but stopped short of dividing the entire region along these lines.[69]

By the time that New France became a royal colony in 1663, the toponyms "Canada" and "New France" were frequently used interchangeably and could refer equally to the nascent settlements on the Saint Lawrence River and the vast region claimed by the French crown. This was a productive conflation for those who imagined French colonialism working on a continental scale.[70] In writing to Jean-Baptiste Colbert, the architect of French colonial policy in the second half of the seventeenth century, the intendant Jean Talon described Canada as "a large country of a vast and prodigious extent" soon after his arrival in 1665.[71] In another letter written the same year, he explained further, writing that "to the North, I do not know the borders as they are so distant from us, and to the south, nothing stops us from carrying the name and the arms of his majesty all the way to Florida."[72] His was a capacious classification, and he was even willing to include distinctive spaces such as Acadia.[73] Talon named Nicolas Denys, a long-time resident of Acadia who held the rights to land between Cape Breton and the mouth of the Saint Lawrence, a "habitant of Canada."[74]

Talon was not alone in conflating Canada and New France. Conceiving of French North America on a continental scale, Pierre Boucher explained in the title to his *Histoire veritable et naturelle* (1663) that "Nouvelle France" was "commonly called Canada." His New France was "a very large country, that is cut in two by a large river named the Saint Lawrence."[75] Jean-Baptiste Franquelin's *Carte d'Amérique du Nord* (1688) directly equated the "Canada or New France" that stretched across the landmass of North America in bold letters.[76] The equivalence was repeated in other continent-spanning maps such as Guillaume Delisle's *Carte de Canada ou de la Nouvelle France* (1703) and Nicolas Bellin's *Partie orientale de la Nouvelle France ou du Canada* (1744), where the title equated the two terms and Canada was drawn across the Saint Lawrence from the north of Montréal to the Bay of Fundy.[77] This ambiguity between the Canada that had been claimed by France in the sixteenth century and the territorial pretensions of the eighteenth-century French crown was maintained in arenas where there was imperial competition for these regions.[78]

Yet locating Canada became a newly urgent task at the end of the seventeenth century, and its spatial footprint retracted considerably. Talon's invocation of a continental New France met with a quick and negative response from Colbert, who, in a 1666 response, explained that "the King cannot at all support your reasoning about the means to make Canada a strong and powerful state."[79] Colbert invoked the king's anxiety about "depopulating" France to the benefit of New France and about the ability to support even those colonists who were being sent across the Atlantic.[80] Soon, however, the spatial reasoning behind what has become known as the "compact colony policy" was foregrounded.[81] A few short years later, Colbert wrote to governor Frontenac, "You must hold to the maxim that it is far more worthwhile to occupy a smaller area and have it well populated than to spread out and have several feeble colonies which could easily be destroyed by all manner of accidents."[82] The policy was embraced by Colbert's successors such as Louis Phélypeaux, Comte de Pontchartrain, who similarly instructed colonial officials to direct their attention to the Saint Lawrence—to Canada—and to the preservation and expansion of the agricultural colony that was taking root there.[83]

These *mémoires* and maps rooted Canada into the settled area of the Saint Lawrence Valley between Montréal and Québec. Although largely separated out into three "islands" of settlement in the 1660s, this region had, by the end of the seventeenth century, become home to the densest French populations in North America.[84] It would become "one continued village" in the eighteenth century.[85] The arrival of Talon and the directed interest of the French state in the affairs of New France effected a remarkable change in the Laurentian colony.[86] This decade saw the arrival and settlement of a number of different communities such as the Carignan-Salières regiment whose soldiers and officers settled at the mouth of the Richelieu and the famous *filles du roi* to whom so many modern Québecois families can trace their origin.[87] Merchants brought over indentured servants, and family farms became socially and financially self-sustaining.[88] The crown invested more than a million livres in supporting the industrial and commercial development of the colony.[89] The effect was the infusion of French ecological and social regimes in what, following Alfred Crosby, we might consider a little "Neo-Europe" on the Saint Lawrence.[90] This focus carved the Laurentian valley out as a region apart from the rest of French North America; it set it apart as Canada.

* * *

Canada grew smaller as the product of a division between the cultivated landscapes of the Saint Lawrence Valley and the forested landscape around them, a division that was read ecologically but that was also registered as a moral, social, and civilizational transition. The official discourse of Colbert and his successors, taken up by the intendants and the governors they sent to rule in their stead, posited a firm division between the *pays d'en haut* and the agricultural seigneuries of the Saint Lawrence Valley.[91] Metropolitan and colonial officials often initially saw success in the *pays d'en haut* as antagonistic to Laurentian settlement but they were formed in a dialectic.[92] The discursive significance of the agricultural lands of Canada was counterbalanced by woods that seemed to lurk at the margins of French settlements and that took root in the imaginations of colonial administrators.

This fissure was most apparently—and has most often been studied as—an ethnic and cultural divide between indigenous and French worlds, but it was most often represented at the time as an ecological transition between cultivated landscapes and an unspecified "woods" (*bois*) beyond them.[93] The woods had long been seen as a home to threatening indigenous peoples. In

1663, for example, instructions sent to a royal investigator warned that if the woods were too close to habitant homes, the Haudenosaunee could "come almost hidden . . . to the homes of the French."[94] Yet these administrators also saw the woods as an avenue for French subjects to travel to Native communities and to learn from indigenous cultures. The writings of colonial administrators during this period make frequent reference to the French coureurs de bois who traded furs in the west, but if it was they who threatened the moral and economic foundation of the colony through their effect on colonial populations and the introduction of indigenous culture, the woods were also figured as a menacing agent in their own right.[95] Minister of the marine Pontchartrain explained at the end of the seventeenth century "that ranging the woods . . . is the source of all the disorders in the colony." Instead, colonists ought to set themselves to "cultivating the land, to fishing, and to other things that he has always recommended and that the nature of the country affords to their application and industry."[96] In 1687, governor Denonville expressed concern about the "disorders and infinite abominations" that followed the marriage of French men to indigenous women.[97] The "commerce" with Native peoples that characterized the lives and livelihoods of the coureurs de bois was therefore understood to be as much cultural as it was economic.[98] French children raised in this climate took on, claimed at least one eighteenth-century observer, the culture of indigenous peoples.[99] The threat that the coureurs de bois posed to Canadian settlement was more imagined than real.[100] Yet, at the time, the woods seemed to threaten the creation of "ensauvagés" colonists.[101]

The culturally and racially mixed communities that would come to be known as Métis were also becoming readily observable, prompting officials throughout French North America to try to limit intermarriage and enforce boundaries that only decades before had been made deliberately porous.[102] Throughout the preceding century, Jesuit missionaries had struggled to come to terms with ambiguous success or outright failure; prominent conversions such as that of Innu Pierre-Antoine Pastedechouan were undone by the inability of converts to live between two worlds, while others such as the Wendat man Amantacha were killed in wars with the Haudenosaunee.[103] "There were intercultural adoptions," writes historian Allan Greer, "but not always in the direction favored by the bureaucrats."[104] A gradual discovery of a more deeply rooted and recalcitrant aboriginal difference meant, then, not only a waning optimism for the possibilities of integration but legitimate fear about the loss of Frenchness among the colonial population.[105]

This was therefore an era where officials, colonists, and missionaries were becoming aware of their limited ability to induce aboriginal peoples to become French. Even where they became French citizens following their baptism, aboriginal cultures remained frustratingly foreign and even frightful, as it often seemed that it was the French who—either by force or choice—moved closer to indigenous cultures and peoples.[106] In 1691, governor Frontenac exclaimed that the way in which missions were being operated often made the *sauvages* "more harmful than useful to the service of the King and even to God."[107] Within a decade, Antoine de la Mothe Cadillac reported that in the debate around his proposed colony at Detroit, the intendant Champigny cast doubt on the entire project of *francisation*. Champigny's complaints, as they are passed to us today, suggested that the "Sauvagesses francisées" by the Ursulines were debauched, "and it is the same thing with the boys that we instruct at the seminary at Québec."[108] Cultural and sexual intercourse were thus conflated and the outcomes of the encounter itself rendered uncertain.

Ecology remained an essential element in these discussions about colonial identities, even if there were fewer evocations of horticultural metaphors or concrete plans to cultivate *sauvage* peoples. In the same way that earlier optimism had thought that settling French and indigenous communities with cultivated plants would support the creation of a New France, colonial administrators wearily accepted that unruly environments could become a destabilizing force in colonial lives. The specific makeup of these woods—what species might have been the most threatening, for example—was left unstated. Instead, the woods were defined by how one lived in them.[109] We might suggest, then, that critics of the woods recognized and feared that humans and the woods became botanical "companion species" in the spaces beyond the cultivated landscapes of the Saint Lawrence Valley.[110] The woods were sites in which French people became subject to "laziness, independence and debauchery" and that doomed those accustomed to it to a life of poverty, wherever they eventually settled.[111] These behaviors threatened the very existence of Laurentian farms. "The clearing of the woods and the cultivation of the soils depends principally in stopping all the young men of the colony who go to trade in distant lands," wrote the intendant Champigny.[112] In the same way that earlier colonists and missionaries understood that changing the ecological relationships of indigenous communities was an essential facet of their civilization, the American woods threatened to transform those French colonists who lived in and traveled through them.

The character of an ideal agricultural region was only vaguely defined in administrative correspondence that most often specified only that it ought to be cultivated.[113] It was enough, for example, for Catalogne to regularly repeat that legumes, grains, and fruit trees grew in French colonial lands, without specifying the identity or origins of the plants in question.[114] Perhaps the clearest articulation of a complete agricultural transformation of the Saint Lawrence Valley was expressed in Sébastien Le Prestre de Vauban's 1699 *Moyen de rétablir nos colonies de l'Amérique et les accroître en peu de temps.* Vauban, one of the leading military engineers of his day, is rightly famous for his work on the fortress at Louisbourg but had a broader interest in French colonialism and, in particular, Canada.[115] Vauban was remarkably ambivalent about Canada's use to France, but colonialism, he explained, was a moral imperative for the civilized world and "without their aid, the land would have taken much longer to fill itself with habitants than it has."[116] European colonialism continued the carefully planned colonial expansion of the Greeks and Romans, and, in this spirit, Vauban laid out a detailed plan for the proper expansion of France's presence in North America that had been, to that point, "languishing and uncertain."[117]

A central facet of his plan was "make the country better known" in advance of actual agricultural expansion.[118] We might consider his plans as much about making Canada legible from Paris or Versailles.[119] This involved careful study of the airs, the waters, the soils, and the general disposition of the landscape such as it would aid or hinder economic exploitation and defense.[120] Vauban's ideal landscape was clearly articulated, as it stretched from defendable heights to vine-covered slopes, to prairies under cultivation, and onto rivers suitable for mills and connection to Atlantic mercantile networks. Each had a clear function and distinct moments of economic and ecological transition. His plan would be directed by the crown, with the labor advanced by "battalions" of officers and soldiers with the skills needed by the colony. Vauban's *mémoire* was particularly detailed about what clearing the land meant. "It is necessary," he wrote, "to remove everything, stones, wood and grasses . . . cleaning fully the land that must be dug up to at least a foot of depth; to leave no trace of roots . . . so that the plow can be passed through freely."[121]

Vauban imagined a wholesale transplantation of European ecologies onto American landscapes that, in preparation for their transformation, were stripped and scrubbed clean of any recalcitrant matter. This was a very different understanding of how cultivation could be used to support New France. The expansion of cultivated space in such a manner was also imagined as an expression of French sovereignty. Pierre Boucher, for example, proclaimed in 1663 that the Haudenosaunee harassed and slowed settlement on the Saint Lawrence and occupied lands to the south that, once conquered, would become the heart of French colonial settlement in the region. The "Country of the Iroquois . . . which is beside our great river," he wrote, "is a very good country & very agreeable: the land is perfectly good & the best that one could find."[122] All that was required for occupation was that "the Iroquois were a little humiliated, or to say it better tamed."[123] Cultivation of agricultural landscapes continued to evoke a peaceful, morally upright colonialism but regularly implied that violence against indigenous peoples and places would nonetheless be necessary.

*   *   *

In the story of late seventeenth-century French North America, the ineluctable western pull of trade and empire often features prominently.[124] Forts, contested almost immediately at their founding by the proponents of a "compact colony," were thrown up at strategic locations throughout the Great Lakes, the Jesuit mission expanded with the explorations of fathers such as Marquette and Allouez to push to the western edges of the region, and merchants sought more favorable markets as the center of the fur trade left Montréal.[125] The slow and uneven dilation of French colonial space into these new reaches provoked serious considerations of its effects. Was it good for trade and the economy of the colony? What effect would it have on the salvation of indigenous people in the region? What would become of spaces and peoples that France claimed as its own? These questions invited consideration of the cultural and geographical status of French colonialism in North America and offered figures such as Catalogne, skilled in the contemporary sciences of space and a prominent voice in discussions about the nature of France's American empire.

Missionaries quickly realized that the defense of western environments was an essential facet of supporting the expansion of their missions, and they worked in their writings to blur the boundaries between woods and cultivated spaces that defined the political ecology of colonial and metropolitan authorities. Against accusations that they needed to focus on bringing indigenous peoples closer to French settlements to succeed, Jesuit missionaries used ecological evidence to argue for the suitability of the Great Lakes region for evangelization and as evidence that French influence could productively move west.[126] Following the explorations of Jacques Marquette and Louis Jolliet in the Great Lakes, the Jesuit Claude Dablon promised that "the soil is so fertile that it yields corn three times a year. It produces, naturally, fruits which are unknown to us and are excellent. Grapes, plums, apples, mulberries, chestnuts, pomegranates, and many others are gathered everywhere, and almost at all times, for winter is only known there by the rains."[127] Maps that Jesuits produced during this period similarly demonstrated the authority of Jesuit observations of the region and reminded the king of his obligation to support evangelization in the region.[128]

Other missionary authors—Sulpicians and the Récollets who had been recruited at the behest of Frontenac—similarly described fertile landscapes for agriculture and evangelization. In 1669, for example, the Sulpician René Bréhant de Galinée reported near what is now Lake Ontario that "in the depth of the woods, one remarks beautiful lands, above all along the rivers that discharge into the lake."[129] That this merited noting seems remarkable, although it is likely that, following the Haudenosaunee conquest of Wendake, the entire area was undergoing a rapid rewilding.[130] Galinée noted that soon after, among the Haudenosaunee, "a number of *sauvages* . . . came to make us little presents of corn, pumpkins, mulberry and blueberries, which are fruits that they have in abundance."[131] Indigenous agriculture gave proof to the charge that there was more to these regions than the woods that undermined settlement and threatened the character of colonists.

Characterizing the environments of French North America soon became an essential tool in debates about the effects of New France's western expansion. Spatial tools seemed particularly apt for colonists well versed in the ecological and climatological complexity of a colony that pushed up against the northern reaches of North America's temperate ecosystems. Pierre Boucher, for example, explained that "the seasons are not equal everywhere in the country: at Trois-Rivières there is almost a month less of winter: at Montréal around six weeks, & among the Iroquois there is only around a month of winter. Québec, even though less favorable for the seasons & for the view of the place that is not very agreeable, has nonetheless a very great advantage because of the number of *Habitans*."[132] Several decades later, Bacqueville de la Potherie seemed surprised that at the Jesuit mission at Chicoutimi "the climate is much harsher than at Québec, even though it is only forty leagues farther up the river." He further noted, as he traveled between Québec and Montréal, that "one remarks that spring commences 15 days or three weeks earlier, one sows seeds there earlier, & the winter also arrives later there."[133]

The difference between these regions was amplified by broader climatic trends. Even as French colonization focused on the Saint Lawrence Valley, the entire region became colder in an era that scholars of historical climates now label a little ice age.[134] Synthesizing the experience of northern North America during two particularly cold periods in the seventeenth century (the Grindelwald Fluctuation and the Maunder Minimum) has proven far more difficult, and the available data suggest dramatic variance across the region.[135] Globally, we can nonetheless point to changes associated with solar cycles and volcanic activity that affected broad swaths of the early modern world. We can equally hypothesize the influence of more localized events such as fluctuations in ocean temperature. The well-known La Niña phenomenon, for example, likely cooled North American climates as Cartier was first arriving in the Saint Lawrence Valley.[136]

While we lack the data to provide specific details about the effects of this little ice age on settlement in the Saint Lawrence, it is certain that it affected local vegetation and that it stressed both the plants that the French painstakingly introduced and those that were at the upper northern reaches of their natural range.[137] Fruit trees, a powerful symbol long mobilized to represent the spread of French domestic ecologies, were particularly susceptible to the climatic extremes of the Saint Lawrence Valley, and the cold was increasingly seen as a primary and immutable characteristic of American environments that threatened their survival.[138] Writing from Québec in 1670, Marie de l'Incarnation complained that "there was still ice in our garden in the month of June." Worse still, she continued, "our grafts of exquisite fruit are dead."[139] Raudot confirmed that imported plants regularly suffered worse against the rigors of the cold. In the early eighteenth century, he explained that colonists had "planted peaches, prunes, cherries and grapes from France; but these trees being too tender to the cold have hardly succeeded."[140] On the other hand, in transplanting Canadian plants into European soils, the French naturalist Henri-Louis Duhamel du Monceau suggested that many should be "as in their country of origin, covered with snow."[141]

The character of colonial society was born, at least in part, from adaptations to Canadian winters.[142] Some promoters sold the cold as a positive for the colony that, as a result, could promise clean air and good health.[143] This was also, however, a very material adaptation. Colonists adopted indigenous technologies such as snowshoes in a bid to adapt to the winter conditions as well as to improve their dress and strategies for household heating.[144] Familiar technologies such as crampons were repurposed to provide footing and safety on ice and snow.[145] Colonists soon realized the wisdom of pulling sleighs along snow and ice in the winter.[146] French preferences for stone buildings gave way to a recognition that wooden houses were less prone to crack in the cold and were easier to keep warm and dry in rural areas.[147] A recognizably Canadian culture was created, at least in part, from these adaptations and their evolution into colonial vernacular forms.

Yet administrative documents also suggest that this adaptation encouraged an official pessimism about the ability of colonialism to effect a continental transformation of ecologies and climates through cultivation. The reality of the cold climate forced considerations of what colonialism could accomplish in the Saint Lawrence Valley and what sort of people could be expected to survive there. It seemed too cold for reproducing more familiar French social and natural worlds or for participating in the new plantation economies that were transforming Caribbean ecologies, economies, and societies.[148] When the intendant Jacques Raudot sought to explain Canada in the eighteenth century, he began with its winter that "commences in this country in the month of October and does not finish until the end of April," during which time "all the rivers, the lakes and even part of the Saint Lawrence River, freezes." Worse still was the snow, which with the cold "produces in this country bad weather unknown in France."[149] Jean-Baptiste Colbert had earlier reported that the cold had threatened the success of the arrival of the *filles du roi* who had been expected to create French families and seemed to jeopardize his plans for the settlement of the colony more generally.[150] In 1689, a "*mémoire* sur les affaires du Canada" both suggested the importation of African slaves and anticipated opposition that these slaves "would die from the cold."[151] We could go on, as the cold figured prominently as a central fact with which to wrestle as authors came to terms with New France.[152]

This more nuanced sense of colonial climates and the climatic variability across New France also produced debates about which environments offered the best possibilities for an agricultural colony. This variability had earlier attracted Jean Talon, who wrote that New France was "a vast country of different elevations, capable in its different climates, and exposures to the sun, of all the productions of ancient France, with no exception. Thus it is hot towards the middle, cold to the north, and temperate in the middle of these two extremes."[153] Yet this sort of climatic equanimity was challenged by others who contested the vision of colonial and metropolitan officials. A late seventeenth-century account of the colony by Lahontan, for example, slipped seamlessly between recognizing the successful settlement of the Laurentian region and calling for western expansion into far superior lands. He explained that "this great country is good or bad following the situation of the climates, the best that is to say the most fertile and the most temperate are up from Quebec towards Louisiana, [and] the four lakes Ontario, Erie, Huron and Illinois that are navigable. . . . The cultivated lands between Quebec and Montreal produce wheat, rye, peas, tobacco, flax, and hemp in a very great abundance."[154] Lahontan later echoed Pierre Boucher's assessment that the Haudenosaunee were "very advantageously situated" and called the Lake Erie region still further south "the most beautiful that I have seen on earth."[155]

Forts built on the southern Great Lakes offered opportunities for agricultural expansion that would not detract from the settlement of the Saint Lawrence. The governor Denonville, for example, reported that the Niagara area offered particular advantages, both strategically against the Haudenosaunee and ecologically. "I am assured," he wrote, "that the lands around [the fort] are very good . . . [and] easy to cultivate. Its situation is around the forty-fourth degree."[156] In 1683, the Récollet father Hennepin suggested that, at Fort Frontenac, "the land along the lake is very fertile, several *arpents* of land have been cultivated, where wheat, vegetables, and kitchen herbs have succeeded well."[157] These were arguments that French ecological lives could be safely extended into land that, at this very same time, others were casting as threatening woods, unsuitable for French culture or cultivation.

The description of these outposts of French settlement bled easily into the charge that these regions were, in fact, superior to the Saint Lawrence Valley. When Hennepin described Fort Frontenac's natural advantages, for example, he also explained that "the winter is almost three months shorter than in Canada; there is reason to believe that a considerable colony will form here."[158] In so doing, then, he expanded the obvious southwestern transition into the *pays d'en haut* and the region being explored by René-Robert Cavelier, Sieur de La Salle, that would come to be called Illinois. It is easy to describe the activities of promoters of western expansion—such as Frontenac, La Salle, or Cadillac—as the product of naked self-interest and to suggest that their outlines of future colonies were chimerical.[159] Nonetheless, as they made their case to colonial and metropolitan officials, they foregrounded complex ecological arguments about the relationships between people and place in North America that alternatively refined and revised the ecological definition of Canada.

Throughout the decade before Hennepin's description of Fort Frontenac, La Salle had been arguing for the need to explore and settle Illinois at least in part through an explicit attack on the environment of Canada. Writing of his experiences between 1679 and 1681, for example, La Salle confidently claimed that the winters in the Great Lakes region were "much shorter than in Canada."[160] Niagara, he explained, had half the winter of Canada.[161] La Salle's descriptions blurred between celebrations of the environments south of the Great Lakes and denunciations of those to the north. He described, for example, "prairies without limit mixed with tall trees, where there are all types of trees with which to build, among other oak, plane like that of France and very different from that of Canada."[162] Accounts suggested, again, that for French colonialism to succeed it would have to find those areas most similar to France. Louisiana was, he argued, more French than New France. His plans for the region would be "better than anything that can be done in Canada."[163] The French who came to this region were following thriving Illinois communities who were already taking advantage of the resources provided by the ecotone transition between forest and prairie.[164] It seems likely that their optimism about Mississippian environments was further stoked by indigenous communities who wanted the French (and their trade networks) to travel further south and to the Gulf of Mexico.[165]

* * *

By the early eighteenth century, environmental knowledge remained an indispensable aspect of knowing American spaces in French North America even as dreams of cultivation withered in the face of ecological experience. At the root of concerns about colonial purpose and imperial identity was a spatial question, asked recently again by two scholars: "Where was New France?"[166] Space was a primary analytic category for actors similarly trying to come to terms with the tensions between extensive and intensive visions of French colonialism. Anxieties about social boundaries were mapped onto geographical imaginaries of colonial space, and environmental knowledge—topography, geology, ecology, and climatology, among others—was mobilized in the same moment that discussions of colonial society invoked nascent human sciences to explain patterns of cultural encounter and exchange.

In effect, New France became more foreign to many colonial and metropolitan observers over the course of the seventeenth century, and the colony seemed less amenable to the sorts of radical transformation that had supported earlier visions of settlement and expansion. The political, social, and economic

challenges that the colony and its inhabitants faced were also epistemological. As wars threatened, conversion suffered, settlement stuttered, and expansion into the heart of the continent was hotly contested, it was not only
knowledge about French colonial North America that was at issue but the
sources of this knowledge and the means through which it was produced. New
France became more difficult to know as wars threatened, climates cooled,
and human and capital costs mounted.

Catalogne's survey of the Saint Lawrence Valley wove together otherwise
seemingly disparate economic, social, and economic threads with a strong
focus on understanding the nature of American environments. During the
remarkable period of transformation in the French Atlantic world after
1663, knowing nature *in* New France became an essential facet of knowing
the nature *of* New France. A century of colonial experience of plants and
places from the Saint Lawrence Valley and throughout the *pays d'en haut* had
educated French colonists, missionaries, merchants, and officials in the extent of American difference. It was experience that, if it failed to result in a
single coherent vision about what empire should be in North America, nonetheless cohered around a method that foregrounded the lived experience of
colonial environments. It was experience that encouraged the formation of
what, to follow Tim Ingold, we might call a "dwelling perspective" that acknowledged that "the forms people build, whether in the imagination or on
the ground, arise within the current of their involved activity, in the specific
relational contexts of their practical engagement with their surroundings."[167]

These tensions remained after 1701, as French imperial officials recognized
the strategic benefit of expansion into the west. Following the urging of
Cadillac, Detroit was founded that year and, growing up and around missions
and indigenous communities on the upper Mississippi, Illinois became a recognizably French space. Efforts at settlement in these areas tempered some of
the early optimism. At Saint Louis, for example, La Salle soon wrote that
"planting is done here only once a year and that is in the month of May, as
there is always a hard freeze in April. It is true that the mildness of the month
of January . . . at first caused us to believe that this country would be as mild
as Provence, but since then we have learned that the winter is not less severe
than that of the Iroquois."[168] Settlement borrowed key elements of Canadian
agriculture such as the organization of land into "long lots" that fronted the
river.[169] Communities that began as missions and fur-trading posts such as
Cahokia (1699) and Kaskaskia (1703) were soon "evolving into agricultural

communities with less transient populations."[170] This region, whether we call in upper Louisiana or Illinois, would quickly become "the wheat granary for the lower Mississippi valley" and thus seems to have fulfilled some of the environmental expectations of La Salle and other early proponents of settlement.[171] It became a unique area of success for frenchification of both people and place.[172]

CHAPTER 5

The Science of Novelty

In 1708, Sébastien Vaillant, a naturalist and botanical demonstrator at the Jardin du Roi in Paris, wrote the first scientific investigation of flora from France's North American colonies. Titled the *Histoire des plantes de Canada*, Vaillant's text emerged from a growing interest in local floras in Europe and the wider world.[1] As much as Vaillant was the author of the text, the *Histoire des plantes de Canada* presented analysis of morphological properties, species distributions, and medicinal properties as authorless matters of fact.[2] When Vaillant described *Alcanna major, latifolia, dentata* (common winterberry), for example, he wrote that "the leaves alternate. It grows to 3 or 4 feet and forms a dense bush. Its branches are full of very little flowers which have the figure of a rosette divided from the center into 6 areas; [it] has 6 stamens supported on a calyx of the same shape."[3] Here and in other instances throughout the *Histoire*, Vaillant followed the generic conventions that had become common in botanical texts that translated the experience of individual plants into a timeless and unchanging flora, removed from the cultural and ecological worlds of French North America.[4]

Accounts such as Vaillant's seem far more familiar to readers today than the sorts of knowledge discussed in preceding chapters. Even if we are uncertain about where or what a calyx is, we can recognize the process of reducing a plant to its morphological parts. Yet the introduction of these methods was revolutionary in their time and announced a dramatic shift in both how and where knowledge about American environments was produced in the eighteenth-century Atlantic world. It is therefore worth noting what is absent in an account such as this as well. There is little sense, for example, of the new means through which colonists were coming to terms with the environments around them: little information was given about colonial or indigenous uses, and neither alternative names nor any ecological context was provided. Vaillant's plants were stripped bare of these features to focus squarely on the morphology of the plant and evidence of its status as a unique and novel species. Figures such as Vaillant—men we are comfortable retrospectively calling scientists—made a powerful argument that it was these novel plants, those that had previously been objects of curiosity valued for their rarity, that were essential for an understanding of New World environments. It was an argument, in effect, that emphasized the newness of New France and that implicitly dismissed the careful study of transatlantic botanical affinities that, to that time, had driven the authors of travel accounts, natural histories, and *mémoires* to present a fundamentally familiar New France to their audiences in Europe. A science that focused on novelty could not support claims to cultivate a *sauvage* New France.

Science became a privileged way of knowing French colonial environments in eighteenth-century North America, in large part because of the Paris-based Académie Royale des Sciences's correspondents in New France. These correspondents were forced to be Janus-faced: responsive to, on the one hand, the environmental and cultural conditions in which they worked, and, on the other, their Parisian patrons who counted among metropolitan administrative and scientific bodies.[5] Their stories open opportunities to investigate the encounter (and collision) of very different ways of knowing the natural world. As these American correspondents attempted to play their limited role within the Académie's networks, they were forced to become mediators rather than collectors, translating the desires of their patrons at the Académie into material rewards and official recognition to create networks of their own that stretched into the interior of the continent. While few traces of their encounters with indigenous and colonial knowledge survive, those that do show that much of what would typify French enlightenment botanical study of North American flora started its life in the encounter of knowledge and peoples at the ecological and intellectual borderlands of France's North American colonies.

It is also through studying this process that we can begin to come to terms with the diminishment of the sorts of voices and perspectives that drove conversations about the political ecology of empire among colonists and missionaries. There was little room in botanical science for the knowledge produced from a dwelling perspective that investigated the cultural and ecological contexts of early American environments. If we are right to be skeptical of the organizational ability of early modern French Atlantic science, we might still consider that, however inefficient bureaucratic French science might have been at producing knowledge, it was demonstrably effective at marginalizing alternative accounts of colonial plants and places.[6] In the annual shipments from west to east across the Atlantic Ocean that contained specimens and narrowly written descriptive accounts, Paris was made a scientific center and Québec became a scientific periphery. In the wake of anxieties about where and under what conditions French colonialism could take root in North America, an emergent botanical science further marginalized calls to cultivate a New France.

* * *

As often as Sébastien Vaillant's manuscript masked the origins of his research specimens, his text still occasionally framed his knowledge of North American flora as the product of a transatlantic conversation with Michel Sarrazin, then the only corresponding member of the Paris-based Académie Royale des Sciences in French North America.[7] The *Alcanna*, for example, was listed as "Envoy de 1703 no 10," hinting at the extra-European origins of the plant.[8] When discussion turned to *Abrotanum* (field sagewort), Vaillant wrote that "M. Sarrazin thought that it was a *Verge dorée* (Canadian goldenrod)."[9] Sarrazin was cited because of the insights he offered into both the natural and cultural worlds of colonial French North America; the references to interviews with "nos Dames *sauvagesses*" complemented Sarrazin's claims to have observed plants such as *Dens Canis flore luteo* (dogtooth violet) both in American soils and in French and aboriginal soups.[10] Distinctions of authorship and authority blurred as Vaillant revealed his dependence on both Sarrazin and the vast American networks of amateur colonial and indigenous collectors who supplied him.[11]

Sarrazin's ambiguous presence in this text highlighted the fundamental tension in the designs of an Académie that sought to focus the circulation of knowledge in the early modern French Atlantic world on Paris but that drew upon the knowledges and collections of an ever-widening array of colonial and indigenous populations to do it.[12] Sarrazin and the other correspondents that the Académie and closely associated Jardin du Roi of Paris sent to the French colonies in North America came to embody this tension between an increasingly centralized academic culture and the need for colonial correspondents who could engage and negotiate with colonial and indigenous collectors. Although their patrons recognized the centrality of the "cultural borderlands" or "biocontact zones" that furnished France with exotic plants, North American correspondents were nonetheless clearly situated at the social and epistemological periphery of the Académie's scientific networks.[13]

The life and work of Michel Sarrazin reveal that he, like the Académie itself, served two masters: science and the state.[14] Bound to the scientific community by election as the correspondent of the botanist and academician Joseph Pitton de Tournefort in 1699, Sarrazin was equally enmeshed in colonial administration, owing much to patronage from the colony's intendants, his place on the colony's superior council, and his position as the royal physician to New France.[15] While in Paris in the mid-1690s, Sarrazin had been able to benefit from the tutelage and patronage of Tournefort, a member of the Académie since 1691 and, then only a few years after the publication of his *Institutiones rei herbariæ*, a hugely influential botanist in Europe and professor at the Jardin du Roi.[16] Sarrazin's botanical research in North America began when his ship first touched land in 1697, after crossing the Atlantic to take up the newly created post of royal physician in New France.[17] Modeled on earlier experiments with royally funded science in the French Caribbean, this conflation of scientific and administrative roles was replicated by Sarrazin's successor to the post of royal physician in New France, Jean-François Gaultier. It was extended to the French colony of Louisiana with the settlement of royal physicians such as Jean Prat and a local apothecary, Alexandre Vielle, who, even if not named official members of the Académie, were closely integrated into the correspondence networks of the Parisian naturalist and academician Bernard de Jussieu.[18] While other collectors—be they clergy, surgeons, or soldiers—would add their efforts to the botanical collections sent from New Orleans and Québec to Paris, it was their dual position as scientists and administrators that privileged the work of correspondents such as Prat, Sarrazin, Gaultier, and Vielle; epistemologically their links to the Académie and its members gave their work a validity that impressed both colonists and officials alike, while their involvement in colonial administration gave them access to the financial and social resources necessary to retain their value to the Académie and its members.

Atlantic communication, if never entirely controlled, was effectively channeled toward Paris and Versailles.[19] Although the establishment of state-centered, bureaucratic institutions was a central means by which colonial correspondents became cogs in a transatlantic machine, recent work has shown the extent to which patronage networks survived and evolved under Louis XIV, who was able to mobilize and influence the individual patronage networks of his secretaries of state.[20] At the same time, stemming from a simultaneous effort to expand France's Atlantic infrastructure through the construction of new ships and the expansion of Atlantic ports, Colbert's reforms gave the French Atlantic a salience that it had previously lacked.[21]

Colbert's efforts to centralize French cultural production had a direct influence on the relationships between the French scientific community and the state. Invested with royal authority after its founding in 1666, the Académie Royale des Sciences quickly became the center of scientific knowledge production in France and, as Colbert and Louis XIV seem to have initially intended, was soon the seat of scientific legitimacy both within France and throughout the French Atlantic.[22] It was invested not only with the authority of Louis XIV's state but with much of its character as well.[23] This meant an increasingly stratified membership, a rich visual culture representing and reinforcing royal authority, and little room for those at the margin of the institution to be anything but informants in networks of epistolary and specimen exchange directed by the Académie or associated institutions such as the Jardin du Roi and the Observatoire royal.[24]

While continuing to fulfill its dual mandate to both foster French scientific advancement and bring glory to the crown, the Académie that Sarrazin entered into was in the midst of a massive reorganization. Reworking both the inner life of the Académie itself and the means by which academicians interacted with the wider world, a reorganization in 1699 codified and reinforced many of the practices that had already become common in the preceding three decades of the Académie's life.[25] A major facet of the reforms involved creating more nuanced membership categories, clarifying the relationship of external correspondents to the Académie's Parisian members.[26] Initially, even as there were clear differentiations in the amount of pay and prestige allotted to particular members of the Académie such as Christiaan Huygens and Giovanni Domenico Cassini, for those admitted there was a rough equality in membership.[27] Rather than being represented by differentiated membership categories, status in the Académie was represented by access to the patrons and protectors of the Académie and carried no official marker.[28] The status of foreigners and correspondents was less clear. Even though there were no official corresponding members before 1699, nonexistent or ambiguous divisions between honorary, corresponding, and standing members, along with the appointment of Jesuits sent to China at the behest of Louis XIV after the 1680s, meant a confused relationship with de facto correspondents.[29] At the same time that these confused categories seemed to grant correspondents an equality with regular members, the Académie's closed sessions reinforced a distinction between its members and the wider French and European scientific community and served as a tangible reminder of its ultimate role as the principal gateway to scientific patronage and legitimacy in France.

Following its official reorganization in 1699, the already hierarchical society became more restrictive still. Stratification increased as membership categories were further refined and multiplied. As full membership in the Académie was further split into subcategories, French religious and amateur scientists were officially limited to the role of honorary members, prestigious to be sure but a clear signal that their future intellectual contributions to the Académie would be limited.[30] Even as a lack of training limited the ability of Jesuits in China to do science in an "academic mode," missionaries—and, indeed, all French religious—were further ostracized as even more categories of membership were created in 1716. While the Jesuit and accomplished astronomer Thomas Gouye had, as an honorary member, served as the Académie's vice president throughout the early eighteenth century, after 1716 religious could only serve the Académie as free associates, ineligible for office and lacking the vote of regular members.[31] Full membership was further restricted to those able to reside in Paris and take part in the biweekly meetings of the Académie that continued into the eighteenth century.[32]

Barred from full membership by their geographical location, for colonial scientists such as Michel Sarrazin the creation of the new membership category of correspondent in 1699 simultaneously ensured access to the prestige and resources of the Académie and an inferior role in its scientific work.[33] While communicating directly with the Académie had previously been difficult or even impossible for colonial naturalists, relationships between colonial correspondents such as Sarrazin and Gaultier and their academic sponsors Tournefort, René Antoine Ferchault de Réaumur, and Henri-Louis Duhamel du Monceau gave colonial botany a regular and attentive metropolitan audience even as it severely circumscribed their role in the scientific networks of the French Atlantic world.

* * *

For the Académie, the placement of official corresponding members and the cultivation of long-term relationships with royal physicians in French colonies in North America was part of a larger strategy to secure reliable and trustworthy sources of both information and specimens.[34] This was an effort that was by no means restricted to either botany or natural history.[35] Extending the reach of Parisian scientific institutions and integrating scientific research into the sinews of the French Atlantic empire, these networked correspondents, it was hoped, would both contribute to the development of French botanical science and develop the botanical resources of American colonies. With the support of Marine officials such as Michel Bégon, who had close ties to France's burgeoning scientific institutions through family and patronage networks, the first generation of state-sponsored and Académie-endorsed naturalists were personally trained by academicians before being sent on foreign expeditions as collectors and observers.[36]

In practice, Parisian naturalists such as Tournefort, Sébastien Vaillant, and Bernard de Jussieu became power brokers in their relationships with colonial naturalists, transforming both their the contributions and rewards into the product of moral obligations to one another.[37] In exchange for specimens and scientific data, correspondents could expect both limited financial compensation and the right to bask in the reflected glory of the Académie.[38] In the letters that Sarrazin exchanged with Tournefort, Vaillant, and Réaumur, botanical specimens were represented as gifts, with no explicit expectation of compensation or recognition; correspondents surrendered control over their discoveries when the ships transporting them to Europe left Québec. Yet the letters that Sarrazin and other North American correspondents sent to their Parisian patrons sought to preserve their connection with the specimens and the written descriptions of local plants that accompanied them. Efforts to increase the perceived value of botanical specimens proceeded through accounts that emphasized the significant sacrifices that were necessary to complete the research that was being freely given.

While the experience of botanist Jean-Louis Guérin was rare—he died of a fever on his first trip to collect plants outside of Mobile in 1737—the sacrifices of colonial science were real.[39] Most correspondents complained openly about the difficult nature of their work. The costs of colonial research, whether in time, health, or money, were frequently discussed in letters sent to Paris and in the letters of intendants and governors who spoke on behalf of colonial correspondents to the Marine and French state. Expressing the frustrations associated with scientific practice in a colony that depended upon canoes and peaceful relationships with indigenous communities for effective communication and transportation, Michel Sarrazin wrote to Réaumur in 1726 to explain that "I do not know if you think that we botanize in Canada as in France. I could cross all of Europe more easily, and with less danger, than I could cross 100 leagues in Canada."[40] French botanists had often traveled abroad as part of their own early careers and recognized the unique physical requirements of botanical research. Tournefort had traveled to the Levant, for example, and knew that "if it is tiring to botanize, it is because it is often necessary to look for plants in the highest mountains, or in frightening precipices, whereas one can learn the other sciences in school or from a cabinet." Even if, as Tournefort continued, "one is compensated enough for this trouble by the pleasure that one has to see the most beautiful part of nature," North American correspondents repeatedly felt the need to adjust the expectations of their Parisian patrons to colonial realities and to valorize their botanical research in the heroic language of scientific explorers.[41]

The transatlantic relationships between French and American naturalists never ceased to mirror wider trends in patron-client relationships that were "vertical, unequal alliances characterized by dependence and by dominance and submission."[42] While Sarrazin styled himself a colleague of his Parisian patrons such as Tournefort and Réaumur, few traces of his scientific research appeared in print during his lifetime. Instead, even as the plants that Sarrazin and other American collectors provided to European naturalists circulated in European networks and appeared in many of the botanical texts of the day, academic botanists in Paris routinely claimed the final authority to name and know North American nature.

The voices of Sarrazin and his colleagues were considered partial and incomplete, only to be cited in their own right when circumstance limited the ability of Parisian patrons to verify and complete their research. In his 1755 *Traité des arbres et arbustes qui se cultivent en France en pleine terre*, Duhamel du Monceau translated the distance that separated New France and Old into a divide between amateur botany and legitimate science, which left correspondents in an ambiguous position. He wrote, for example, that "uneducated travelers have often spoken to us of the *Merisier de Canada* (Wild Cherry) that is very different than that of Europe; but we have never been able to get a precise idea of this tree, but now with the seeds that have been sent to us from the same country, we have recognized that the tree that is called the *Merisier* in Canada, is in reality a Birch with the leaves of a *Merisier*. Similarly, the Canadians have often represented the Bonduc as a type of *Noyer*, but the botanists have given us an idea much more exact through the methodical description that they have made."[43] Duhamel du Monceau, who was often more generous in his praise of his North American correspondents than were other French naturalists, nonetheless disassembled the testimony of his correspondents, referencing only those facets that he was unable to supplant with his own

experiences and observations. When he described a Canadian maple, for example, he referenced Gaultier's accounts of Canadian winters but prioritized
his own observations otherwise.[44] Sarrazin's description of the *Assiminier*, a
small tree that grew in Canada and along the Mississippi, was cited simply
because the plant had failed to grow in Europe. "All that we can know of this
tree, we have from several Voyageurs, and in particular M. Sarasin, Royal
Physician in Canada and from M. de Fontenet, Royal Physician in Louisiana.
We do not know if it has grown yet in France."[45] Most often, however, Sarrazin
and his American colleagues were victims of the passive voice. When, as
Duhamel du Monceau suggested, it was true that plants and seeds "had come
from Canada," there was little credit to accord the North American naturalists who had sent them or who likely had studied them themselves.[46]

While part of this was due to Sarrazin's inferior role as a corresponding
member of the Académie, it also reflected several assumptions about the proper
means through which to determine the essential character of a plant and strip
away merely accidental morphological features that clouded proper systematic botany. If, as the botanist Sébastien Vaillant suggested, plants "are produced by Nature following an always simple and constant law, they have
everywhere the same structure, from which they never distance themselves,"
it was nonetheless understood that these morphological patterns were not visible to just any observer in any location.[47] As astronomical data depended on
the placement of observers the world over, so too did botany depend on multiple observations of the plant from multiple environments, most often in botanical gardens in France and Europe that were involved in seed and plant
exchanges with botanists at the Jardin du Roi in Paris.[48]

These "big sciences" privileged the central role of the compilers and ana-
lyzers of these disparate and diverse observations.[49] Through their extensive
gardens, herbaria, books, and correspondence they had at their disposal at the
Jardin du Roi and for the simple fact that they were in Paris, naturalists such
as Vaillant and Tournefort could claim to reorder and recast knowledge of
flora the world over.[50] To that end, while Sarrazin sent his own observations
to his Parisian correspondents, his role was principally focused on sending
specimens of dried plants for his patrons' herbaria as well as seeds and living
plants. Although Sarrazin and the other correspondents of the Académie in
French North America were epistemologically privileged in a colonial setting,
they remained outlying nodes in a network centered on Paris.

* * *

Sarrazin acquired credit with his Parisian patrons by regularly furnishing them with rare or previously undiscovered plants. While never divorced from the search for new botanical commodities, colonial correspondents responded to an academic culture that prioritized the identification of novel plants and the elaboration of taxonomic schemas. In his 1693 *Elemens de botanique*, Tournefort made his own—and by extension the Académie's—priorities clear when he wrote that "knowledge of the virtues of plants, which constitutes the second part of botany, is without comparison more useful than the first; but knowledge of the names of plants must necessarily precede that of their virtues."[51] The pursuit of novelty lay at the heart of the relationship between the naturalists of the Académie and their North American correspondents. The valorization of new plants deeply influenced scientific practice in French North America as colonial naturalists sought to appeal to and attract the support of their Parisian patrons. Yet novelty always remained relative; what was new and therefore valuable for one party to the exchange was old and redundant to the other. As in the scientific networks of eighteenth-century Europe, the economy of botanical exchange was therefore centered on the exchange of "unlike-for-unlike."[52]

In the hunt for novel flora, correspondents and royal physicians presented several distinct advantages for Parisian naturalists. This included at least some preliminary training in medical botany that set them apart from other would-be informants. Sarrazin's experience with Tournefort, who became as much a teacher and patron as correspondent, would be repeated by Gaultier, his successor at Québec, who was vetted by his correspondent Duhamel du Monceau. Practically, they learned the methods of botanical identification of Tournefort and, later, of Jussieu and were given firsthand experience of the Parisian botanical collections, knowledge of definite use when deciding whether a foreign plant was already known in Paris and therefore whether it was worth sending.[53] While both Bernard de Jussieu and Duhamel du Monceau wrote guides for their foreign correspondents outlining the procedures for collecting and preserving plants, the Parisian experience of American correspondents made them invaluable resources and elevated them above the rank of simple informants in the Académie's Atlantic networks.[54] More importantly, however, as a result of their training, these American correspondents not only were reliable informants for their Parisian patrons but entered the networks of the Académie as concretized examples of the scientific personae upon which French botany depended.[55] Equipped with the social and scientific skill set required to expand the reach of French naturalists and the Académie, they simultaneously shifted the power that previously lay with geographically privileged colonists in favor of the Académie and its members.

Académie-trained naturalists such as Sarrazin brought with them not simply a particular skill set but a moral economy of observation that privileged close attention to visible natural phenomena as a legitimate source of knowledge.[56] Tournefort, Sarrazin's first academic patron, wrote in his hugely influential *Elemens de botanique* that "the study of plants does not fatigue the imagination much when one proceeds with method. Their figures present themselves easily to the mind, when one is accustomed to observe them by their essential parts."[57] One had to be educated to recognize those "essential parts" by which to study a plant, to be sure, but the descriptions of North American flora that naturalists such as Sarrazin prepared for the Académie were authenticated as much by their method of observation as by the technical aspects of their work. The truth was in the details, and it was the habits of these correspondents—their patience, their attention to the minute morphological features of their quarry, and the precision of their language—that made them invaluable and trustworthy sources.

The research of French correspondents was often explicitly directed toward discovering the possible utility of American flora, and they were often counted upon to provide expert advice to Atlantic administrators. Jean-François Gaultier, for example, was directed toward investigating the forests of New France by his Parisian correspondent, Duhamel du Monceau. Looking to Canada in his effort to improve the French navy, Duhamel du Monceau requested that Gaultier investigate the production of tar and other shipbuilding materials, providing a detailed questionnaire and instructions to guide both the production and recording of the results.[58] The only time that Michel Sarrazin's botanical work was published by the Académie, it dealt explicitly with the process of obtaining sugary syrup from Canadian maples.[59] For metropolitan naturalists and authorities alike, the trained correspondents of the Académie provided important technical expertise.

Yet the experience of these colonial naturalists was particularly important in establishing the novelty of American plants that at least superficially seemed familiar. When he described a Canadian fir, for example, Gaultier wrote that "one counts two species of them which at the first view are very similar."[60] Subjecting the fir to a more piercing gaze than might amateur naturalists, Gaultier asserted that in spite of the considerable similarities in floral and vegetative characteristics, the species could be distinguished by differences in the color of their resin.[61] Vaillant warned that local climatological differences could produce variations in plant morphology that could fool any but the most astute botanist. Describing the changes that could take hold once a plant had been transplanted in a French garden, for example, he wrote, "We could compare them to a harvester, who broken from labor, sunburnt, and accustomed to eating black bread and drinking water, was to be transported to an effeminate court, where he lived in idleness and softness."[62] While the novelty of many plants from North America was evident at first sight, the fine-grained analyses that established the novelty of many North American plants were opportunities to demonstrate the attentive gaze of the trained naturalist as much as they were statements about the flora of northeastern North America.

The value placed on novelty meant that colonial botanizing resembled less a modern biodiversity survey than a cataloguing of North American plants. The botanizing surveys of correspondents such as Sarrazin put little value in place, focusing instead on amassing a complete collection of plants that grew anywhere in France's American colonies.[63] While only a traveler to New France himself, Pehr Kalm, a Swedish botanist and student of Carolus Linnaeus, left the best accounts of what it meant to botanize in eighteenth-century New France. Having trained with Linnaeus at the University of Uppsala, Kalm had been sent across the Atlantic to collect new and potentially valuable plants.[64] At Linnaeus's behest, he traveled between Philadelphia and Québec between 1748 and 1751. His records of that travel provide an unparalleled view of both the human and natural histories of British and French North America.[65] At each stop, the Swedish botanist would set out into the nearby countryside and collect the plants that had been requested by Linnaeus and those that seemed to him novel, rare, or potentially valuable.[66] The possible utility of American plants remained a consistent focus of Kalm's research, but his text made his overriding concern for novelty explicit. Describing his search near Québec, for example, he lamented that "I took myself next towards a group of beautiful prairies which had the appearance of magnificent copses of leafy trees; but I discovered nothing I had not seen before."[67] The novelty of plants such as Gaultheria (*Gaultheria procumbens*, named for Jean-François Gaultier by Kalm and Linnaeus) was summed up in technical descriptions that stretched entire pages and that aimed to translate his experience of each morphological feature for his European audience.[68] When Kalm did occasionally make a second reference to a plant that he had already discovered, it was cursory. Near Québec, for example, he wrote: "Veronica. Plant described when I was at *fort St. Fred.* located here in several humid places," suggesting that any further description was unnecessary to his purposes.[69]

It was through these practices that novelty was identified, and it would soon become a currency for colonial naturalists who were far more interested in cultivating transatlantic ties with Paris-based patrons than they were a colony in North America. From Paris, these plants circulated throughout Europe, further distancing the source of these collections and obscuring their origins in cross-cultural encounters in the heart of North America.[70]

*   *   *

An older historiography of American science has often attempted to portray colonial botanists such as Sarrazin as solitary practitioners single-handedly cataloguing ecosystems and packing them up for analysis by their grateful French counterparts. They are often described as having introduced modern science and medicine into the peripheries of the Western world.[71] However, the life and work of Michel Sarrazin demonstrates that there was a finite limit to the work that could be done by any American correspondent, let alone those who by virtue of their status were simultaneously expected to treat the sick, manage their properties and families, maintain good relations with their patrons and superiors, and help govern the colonies as members of administrative councils.[72] Sarrazin and Gaultier were intended neither to collect solely on their own nor to navigate and negotiate the extreme distances involved in collecting specimens in the interior of the continent. Rather, from the beginning of Colbert's efforts to reorganize scientific communication, the Académie's interest was in securing trustworthy and reliable sources whose reach extended into the interior of the colony and who could function as patrons and power brokers in their own right. At the same time that the Académie extended scientific networks across the Atlantic and into the North American continent, they became dependent upon the ability of their correspondents to manage networks of their own, drawing upon both the efforts and knowledges of the colonial and indigenous populations of French North America.

Firmly established in the hierarchy of the Académie that left little doubt about their epistemic inferiority to regular Paris-based members, colonial correspondents such as Sarrazin might seem simply a tangible example of what Ralph Bauer has referred to as "epistemic mercantilism," where the production of knowledge was limited to metropolitan centers.[73] Yet far from simply being reduced to the role of informants, these correspondents in fact created polycentric, if still stratified, Atlantic networks of scientific communication.[74] Established in Québec and New Orleans, correspondents of the Académie reproduced the relationship between France and its American colonies in their relationship with the colonial informants on which they drew. At the same time, however, they reproduced the tensions inherent in a system of exchange that, while dependent on colonial and foreign collectors, rendered them invisible in academic networks. These naturalists were, like other arms of the colonial presence in French North America, "simultaneously coerced and coercing."[75] As a result, even as Sarrazin and his colleagues catalogued the flora surrounding the settled areas of New France, Québec was being fashioned into a botanical entrepôt that drew the collections and translated the knowledge of colonial and indigenous plant collectors and readied them for transport to France as scientific specimens.

As local officials and academic correspondents, colonial naturalists were able to draw upon resources unavailable to other local collectors and would-be correspondents to expand their reach into the interior of the continent. Within a decade of Sarrazin's arrival in New France as the first official American correspondent of the Académie, the ministry of the Marine was making efforts to integrate his research into the military networks of the colony. In 1707, the minister of the Marine Jérôme Phélypeaux, the Comte de Pontchartrain, wrote to New France's intendant Raudot, himself a supporter of Sarrazin's work, ordering the "Officers of the King . . . to receive these cases of plants into their buildings without difficulty, to have them placed in an appropriate place where they will be preserved and to send them to my address."[76] Even when Sarrazin departed on his own expeditions, his integration into these administrative networks meant that he was never alone. In the same year that Pontchartrain wrote the letter just mentioned, Raudot and governor Philippe de Rigaud Vaudreuil informed their superior that they had provided Sarrazin with both voyageurs and a canoe to facilitate his collection of specimens.[77] Combining his own research with the fruits of the labors of local collectors and workers who became "invisible technicians" in written accounts, Sarrazin and those who would follow him rationalized collection and gave the flora of North America an official voice, even if it was one that remained unequal with those in Paris.[78]

Records of correspondence preserved in France's colonial archives show that, within a few decades, academic correspondents and scientifically inclined administrators could draw upon a network of post commanders, medical personnel, clergy, and the occasional colonist. In 1736, the governor of New France, Charles de la Boische, Marquis de Beauharnois, wrote to the minister of the Marine, the Comte de Maurepas, assuring him that he had forwarded the minister's request for collection of a particular type of wood to the Sieur de Noy-elle, post commander of Detroit. In the same letter he also praised the com-mander of Fort St. Joseph, Jacques-Pierre Daneau de Muy, author of a *mémoire* on medicinal plants of the Great Lakes that the intendant was also forwarding to his superiors in France.[79] The Académie's correspondents could serve as bro-kers in these exchanges, collecting specimens and translating written accounts for an academic audience. In 1749, Gaultier received botanical specimens from the military officer Gaspard-Joseph Chaussegros de Léry, who had only re-cently returned from a trip to Detroit. At the same time that Chaussegros de Léry's maps went to military officials, Gaultier was entrusted with seeds and specimens that he was to preserve and ready for shipment to France.[80]

Pehr Kalm was clearly impressed with both the reach and efficacy of the network that centered in Québec on Gaultier and then governor Roland-Michel Barrin de la Galissonnière. The ease with which he was able to study soils from the interior of the continent and his evident satisfaction with the botanical collections of Québec testified to the successful creation of a bo-tanical entrepôt in New France that integrated royal resources and academic methods.[81] The governor had also requested that Gaultier produce a *mémoire* "on the means to search for and discover all the trees, the plants, the animals, the types of rocks, the minerals," which was then sent to post commanders throughout the colony. Kalm wrote that de la Galissonnière had promised that "the people who made an original discovery were paid and had their names noted to remember them at the occasion of a vacancy of a post."[82]

Québec became an obligatory point of passage between networks of con-tinental and Atlantic exchange; it became a botanical way station, akin to the royal gardens in Atlantic port cities such as Nantes and La Rochelle where foreign plants would be resuscitated before being sent on to Paris.[83] While even this would not be enough at times, with whole shipments of plants such as those sent in 1717 dying on their trip across the Atlantic, the conflation of administrative and scientific communication was a surprisingly successful effort to mitigate the difficulties imposed by the distance between France and its American colonies.[84]

Sarrazin and his colleagues repeatedly demonstrated their commitment to the Académie's dismissal of materialist motivations for scientific research. Even if their official position as royal physicians made their scientific work obligatory and even as they received payment throughout their work as colo-nial correspondents, the performance of disinterestedness functioned as a marker of epistemological legitimacy.[85] Sarrazin's requests for compensation for his research expenses were put to minister of the Marine Pontchartrain on his behalf by the intendant Raudot; eventually he was rewarded with a payment of 500 livres annually for his "zeal for the research of plants and animals" that he had demonstrated in years of service.[86] In 1749, Gaultier complained politely but directly to minister of the Marine Maurepas about eight years in which he had failed to receive compensation for his expenses, asking for land and fishing rights in lieu of money.[87] In his correspondence with the academician Réaumur, Gaultier similarly emphasized the expenses of colonial natural history and thanked his patron for encouraging the minister to extend the gratification that he felt was his due for his years of service to the Académie and the crown.[88]

As the reach of the Académie extended into North America, the relationship between Parisian naturalists, their colonial correspondents, and amateur collectors became increasingly strained. Colonial botany in French North America was rarely a selfless pursuit. Galissonnière, both governor of New France and a respected naturalist in his own right, criticized the commitment of French colonists to botany and the natural sciences in a letter to Duhamel du Monceau. In New France, he wrote, there was none of the selfless support of discovery on which science depended. Instead, self-interest reigned in a colony where naturalists such as himself and Jean-François Gaultier found few people willing to help them.[89] Colonial collectors repeatedly demonstrated their inability to operate in the ostensibly disinterested relationships that governed the scientific exchange of plants. Gaultier lamented in 1753, "I would have wished that the collection and this shipment had been more considerable. But that has not been possible because of my correspondents who did not keep their word."[90] Situated in an Atlantic economy that traded collections for the favor of the state and the Académie, colonial naturalists were able to receive payment for their work without jeopardizing their privileged position as disinterested scientists but only as they referenced their service and commitment to their patrons and the crown, a move that simultaneously naturalized a status inferior to the Académie and the wider French scientific community.

*   *   *

In effect the North American correspondents of the Académie became mediators between divergent botanical economies, transitioning plants between the networks of the Académie in which plants drew their value from their novelty and were offered freely as gifts and a colonial economy in which plants were acquired from indigenous and colonial collectors based on the commercial values of botanical commodities and the physical labor of transport and collection.[91] Amateur collectors were therefore as suspect for their motivations as they were for their methods and were understood to be a means to an end by the Académie and affiliated naturalists. Their contributions were effaced as their specimens changed hands and crossed the Atlantic, and they only appeared in surviving correspondence as excuses for failure to satisfy the Académie's demand for specimens.

As collections passed from amateur collectors to Québec and the Académie, and as they entered the parallel transition from manuscript to printed text in Paris, the diversity of the continental networks upon which colonial naturalists drew was effaced. While the networks of Sarrazin were made untraceable in the *Histoire des plantes du Canada*, the best picture of the sorts of resources that colonial botanists were dependent upon comes from the writings of Pehr Kalm. Kalm claimed that "not a single botanist had yet researched or carefully described the plants that are found there," yet working routinely with Gaultier and drawing upon the networks of Galissonnière, his text revealed the existence of French networks of plant collection upon which his own work would ultimately depend.[92] Aboriginal peoples appear in Kalm's text as objects of study more often than as informants, and his reliance on local soldiers and colonists is apparent even if his references to a generic "les Français" are frequent.[93] Naming informants such as a certain monsieur Cartier of Cap aux Oies or Father Coquart of Baye St. Paul was partly, he tells us, a strategy to stave off possible critiques, reminding his reader that not all of the opinions expressed in his writings were his own.[94] The information and specimens that these sorts of informants provided, however, were vital to the ultimate success of his work. This was based on his specific interest in finding and collecting useful and potentially valuable plants. By interrogating local residents to find out which plants they identified as noteworthy, Kalm was, for instance, able to learn about sugar maples from a monsieur Chambon at Sault-au-Récollet.[95] His reliance on local informants was also, however, a product of the nature and timing of his expedition. With an itinerary that meant he had often missed the appropriate season for fruits and flowers by the time he arrived, local informants became crucial to the success of his work. Relying on locals who could connect their recollections of flower types with the plants around him meant that he could, in effect, collect plants out of season that might otherwise have escaped his attention.

Kalm's investigations into the natural environments of the Saint Lawrence Valley and the eastern Great Lakes were unique only insofar as he publicly credited informants that in the work of French naturalists such as Sarrazin and Gaultier frequently remained anonymous. Indeed, while the respected correspondents of Louisiana such as Prat and Vielle proved that an official position within the Académie was not essential to becoming a long-term correspondent with its members and the Jardin du Roi, the reality is that most of the collectors who existed outside of the administrative and scientific apparatus of France's North American colonies and the wider Atlantic world became anonymous in texts that clearly depended upon their existence. In Sarrazin's work (as it comes to us via Sébastien Vaillant and the *Histoire des plantes de Canada*), amateur collectors in any part of the colony are virtually impossible to locate even if it is evident that they were essential to his success as a colonial botanist. Referring generally to the practice of colonists or indigenous peoples, for example, the contributions of specific people were rendered invisible. Writing about the *Sanguinaire* (Bloodroot) that he sent in a shipment in 1698, he recounted that "I have been assured that they provoke menstruation" by indigenous women.[96] In moments such as these, he hinted at the existence of a vernacular colonial knowledge and potential sources of physical specimens, but he nonetheless refused to identify a specific source or credit the information beyond recounting its existence. Instead, Sarrazin also wrote simply that "I do not believe" these aboriginal claims, further distancing and dismissing the legitimacy of indigenous contributions to French botanical science.[97]

This practice of erasing the contributions of the local collectors upon which colonial correspondents depended continued in the work of Jean-François Gaultier, Sarrazin's successor at Québec. The descriptions in his manuscript *Description de plusieurs plantes du Canada* suggested that the simple fact that such correspondents were themselves long-term residents lessened reliance on local informants. While the need for knowledge of particular morphological features such as flowers, fruit, and leaves made Kalm reliant on informed locals, Gaultier's discussion of *Abies conis sursum sive mas* (a species of fir), for example, testified to his year-round observation of a plant that could have misled an itinerant naturalist as it changed its appearance in successive months and seasons.[98] Yet in his descriptions of plants that included medicinal and other uses, it is clear that he was forced to cast a wider net. Adding to Sarrazin's description of *Arum Canadense foliis ad Betam accedentibus* (Skunk Cabbage), for instance, he wrote that "the habitants of Canada employ the root of this *arum* for the flux with children."[99] As with Sarrazin, Gaultier's use of pronouns such as "on" hid his reliance on specific sources of information and effaced the ethnic, gender, and geographic contexts of much of the knowledge that he collected and presented to his patrons at the Académie. Even where a specific source was mentioned, the details provided attested more to an anxiety about justifying the information to a Parisian audience than an effort to introduce colonial collectors into French scientific networks. When describing the medical use of a colonial maple, for example, he cited "a Surgeon from the Isle d'Orleans," clear evidence that a degree of competency associated with medical education and training mattered as much for the validity of colonial science as any ethnic, gendered, or social distinctions.[100] The result was that while colonial botanists frequently named European contemporaries and the sources of their taxonomic conventions such as Bauhin, Gronovius, and Linnaeus, aboriginal and colonial populations figured in a generic sense as implicitly male habitants and *indiens* and never as sources of knowledge in their own right.[101]

Information on some rather exceptional collectors does survive in the archives of administrative correspondence between New France and Old, but these references are rare and the typicality of these particular collectors is difficult to ascertain. Tellingly, these collectors appear in documentation today because they sought to skirt the Académie's trained correspondents and contribute directly to both the Académie and the French state. In the document series C11A that contains letters sent from colonial Canada to France, for example, it is possible to find reference to only eight named collectors, three who were in the military, three who were colonial clergy, a solitary widow, and the colonial physician Hubert-Joseph de la Croix, who, after Sarrazin's death, had hoped—for naught in the end—that collecting plants for the Jardin du Roi might make him an attractive candidate for royal physician.[102] While de la Croix's interest in providing the Académie with plants is clear, the motivation of many of these other collectors remains an open question.[103]

Attempting to bypass correspondents such as Sarrazin, these amateur collectors nonetheless remained dependent on the resources of the state and, more specifically, the infrastructure of the French Marine. In almost every one of these cases, whether it be the three boxes of plants that the Abbé Jean-Baptiste Gosselin sent in 1738 or the "packet" sent by the widow Lepaillieur in 1740, would-be collectors were reliant on the royal ships to transport their plants and on the network of gardens and correspondents that connected port cities such as Nantes, La Rochelle, and Rochefort to Paris's Jardin du Roi.[104] At least in part, this was likely a recognition that the efforts to integrate scientific and administrative networks under way since the late seventeenth century had succeeded and that plants that were transported on state-owned ships were more likely to arrive in France in a condition that would please potential patrons and leave them scientifically useful.[105]

Relying on state-financed and state-supported networks was not always feasible or possible, and even making use of the royal ships was no guarantee that plants would survive their transatlantic trip. In a letter written in New Orleans in 1737, for instance, Prat asked his patron Bernard de Jussieu to ensure that "the commandants of both merchant and royal vessels had orders from the minister to accept and to take care of all of the plants that I give them for the Jardin du Roi," adding that it was hard to see the efforts that he had spent to gather the plants go to waste as his collections arrived in France damaged, destroyed, or dead.[106] Yet with few other alternatives, colonial plant collectors such as Gosselin and the Québec-based physician de la Croix made use of the networks established by the Académie and French Marine. It was in these routine submissions that the colonies were made a scientific periphery; legitimating the "epistemic arrogance" of the Académie as they loaded their plants aboard the ships of the Marine, they accepted a role as collectors and informants and gave Parisian naturalists the capacity to speak about North American flora in their stead.[107] Prat's complaint, however, revealed that the Académie remained forced to continually divert botanical specimens from networks where they were defined and valued by their marketability as commodities, both at land in North America and at sea in the Atlantic world.

The best records of the activities of individual collectors who sought to integrate themselves in French scientific networks are those that note their compensation for expenses incurred as they collected plants and other natural historical materials such as minerals and animal products. Yet it is clear from the reports of their expenses that these collectors relied on the labor of local colonial collectors as much as the Académie's correspondents did. The Abbé Gosselin, for instance, was compensated for 199 livres "that he had spent in his research" for the year 1738 and, though the amount required grew to 250 livres, he was compensated again in 1739.[108] As items such as "five days labor of two canoers" that were on his list of his expenses make clear, most of the costs incurred by colonial collectors such as Gosselin were related to transport of both himself and the plants that he was able to gather. Paying for provisions and the wages of the men who paddled for him, his expenses even included the costs of the cases in which the plants would eventually be sent to France.[109] Gosselin was also cryptically compensated for twenty-three livres "paid to the *sauvages* for some new plants" near Beauport, hinting that colonial collectors such as Gosselin could be crucial links between Parisian naturalists and aboriginal peoples but providing precious little detail about the sorts of negotiations that this type of contact involved.[110] As the *Dictionary of Canadian Biography* explains, "Although certain historians have called Gosselin a renowned botanist of the period, he deserves rather the title of plant collector."[111] The reality is thus that, in a global network of plant collection, there was little beyond specimens that collectors such as Gosselin were permitted to contribute and for which they were (and continue to be) recognized.

The Jesuit Joseph-François Lafitau presented a rare case where a colonial naturalist who sought to present his own specimens and natural historical research to an elite audience in France was able to do so.[112] Sending a written account of his discovery to his confrères in the Society of Jesus in Paris in 1717, an excerpt that featured his ethnobotanical findings was published the same year in the Jesuit *Journal de Trévoux*.[113] This account was read by the Académie Royale des Sciences that same year, and it was his reliance on indigenous sources that was deemed particularly noteworthy even if many present objected to its findings. Defending his work in his 1718 *Mémoire*, Lafitau's response to the criticisms leveled by the Académie demonstrates how the Jesuit understood his place in the French scientific community.[114] In providing specimens to both the regent of France and the Jardin du Roi, printing his own account that silenced Sarrazin and the scientific apparatus that the Académie had extended into the French Atlantic world, and daring to suggest that Bernard de Jussieu and Sébastien Vaillant were epistemic equals in print, Lafitau's experience highlights the plight of collectors such as Gosselin and de la Croix and the difficulty of contributing to French botanical science without going through the Académie's American entrepôt and the colonial correspondents who managed them. Lafitau's fault was that he wanted to contribute to French science without recognizing the role of the Académie Royale des Sciences as arbiter. Just as fellow members of his society in China discovered as they sent the Académie silkworms and other pieces of naturalia, there was little room for outsiders to be anything but providers of specimens, particularly for those who, like Lafitau, sought to shift the center of French scientific networks to cultural and ecological borderlands of the Americas.[115]

Even if colonial collectors were experts in the sorts of negotiations with aboriginal peoples that historians such as Londa Schiebinger suggests were vital to the success of colonial botany, the content of their scientific practice remains largely unknown.[116] However, thanks to a single *mémoire* that survives in the Muséum national d'Histoire naturelle, it is possible to demonstrate that at least some of these collectors agitated for a greater role than they ultimately received in the scientific production of the Académie Royale des Sciences and in the manuscripts of colonial botanists such as Prat and Gaultier. The manuscript titled "Mémoire sur les plantes qui sont dans dans le caise. B" appears to have originated at Île Royale (now Cape Breton) around 1725 and was likely written by François-Madeleine Vallée, an engineer recently exiled to the colony by order of the king (Figure 9).[117] It is part of what would have been a number of texts that were included in the boxes of plants sent to France and the Jardin du Roi by anonymous colonial collectors in French North America in the eighteenth century. To my knowledge this is the only document of its type to survive, and the rarity of these texts testifies to the limited role envisaged for colonial collectors by the Parisian naturalists who received their specimens and the success with which the contributions of such collectors were ultimately erased.

The text itself is poorly written, misspellings abound, and, in general, it is illustrative of the problems that the creation of regular correspondents in American colonies were meant to solve. While the criteria by which botanical texts organized their descriptions of plants remained contentious throughout the early and mid-eighteenth century, this text, which is neither alphabetical nor organized around any of the visible characteristics of the plants discussed, displays a structure whose rationale was likely only visible to its author. The plants, from *Sarrazine* (Purple Pitcher-Plant) to *Thysaouyarde* and *herbe à jean hébert* (Bloodroot), seem a hodgepodge most likely included together because of an ecological relationship or a common collection location; aquatic plants figure prominently, and it seems likely that many were collected along the coasts and in the swamps near Louisbourg.[118] For academic readers at the Jardin du Roi, the principal problem with the text would have been the inability or unwillingness of the author to situate the plants in taxonomic conventions that affixed and stabilized botanical identity on both sides of the Atlantic. Instead of drawing upon Tournefort or Vaillant to name these plants, the author instead relied upon vernacular and indigenous nomenclature that, while sometimes included in the work of Sarrazin and Gaultier, was absent in the published works of Parisian naturalists.[119]

The ecological and medical knowledge of both aboriginal peoples and French colonists figured prominently in this text that, while addressed to French naturalists, ultimately aimed to subvert the relationship envisaged by the Académie. Even where indigenous knowledge was not explicitly cited, it was assumed that these peoples might have had something to add to the account. "The *sauvages* say nothing of this plant," the author wrote about an

Figure 9. Herbe à Jean Hébert, Laboratoire de phanérogamie, Muséum national d'Histoire naturelle, Paris. Photo by author.

unnamed plant that he nonetheless hoped his Parisian readers could identify, for instance.[120] While dependent upon the plants included in "caise B," readers such as Bernard de Jussieu and Sébastien Vaillant would have been unwilling to accept that the identifications of the "creoles" or *sauvages* were a legitimate contribution to botanical science, and, as these sorts of documents were neither preserved nor included as sources in published botanical texts, it is clear that they eventually disappeared. Yet, while dependent on state resources for the transport of their collections, it is also clear that at least some colonists imagined a far greater role for themselves in the production of knowledge about North American flora than Parisian naturalists ultimately permitted them.

This text, while providing evidence that colonial collectors were active in seeking to furnish Parisian naturalists with botanical specimens, is similarly effective at demonstrating the ease with which the participation of indigenous and colonial populations in French scientific networks was effaced. Indeed, it is almost solely through the glimpses provided by texts such as that above and the indexed expenses of Gosselin that it is possible to see how aboriginal peoples participated in continental and Atlantic networks. While colonial correspondence and travel accounts amply demonstrate, for example, that aboriginal people became key furnishers of plants such as American ginseng once commercial opportunities had been discovered, their role in producing scientific specimens is difficult to ascertain.[121] The specimens that survive in the Muséum nationale d'Histoire naturelle provide little fresh insight. Even where there is evidence of an interest in indigenous or colonial usage, details on the acquisition of ginseng and other plants are scarce. For instance, while Prat's successor to the title of royal physician in Louisiana, Bénigne de Fontenette, provided ethnographic data on the usage and indigenous names of American ginseng that were included with the physical specimen, this information served to cloud the origins of these specific plants rather than illuminate them. Although he attributed his information to the Natchez and the Illinois, any specific aboriginal communities involved in the collection of the specimens were effaced.[122] Looking at the specimens preserved in France's national herbarium today it is clear that this information was considered supplementary and was included to hint at potential medicinal properties (snakebites and wounds are mentioned) rather than to credit indigenous cultures with a unique insight into the plant's nature.[123] Yet even this level of detail is rare in an archive that is devoted principally to the preservation of physical specimens that have been abstracted from specific cultural and ecological contexts.

This echoes the treatment of indigenous knowledge and peoples in the extant writings of Sarrazin, Gaultier, and Vielle. Sarrazin treated indigenous sources in entirely the same manner as he did colonists, which is to say that he summarily dismissed them. Similarly, Gaultier's writings often make it impossible to distinguish whether the source of information about the uses of a North American plant came from a colonial or indigenous community, and both were regarded with a cultivated detachment. When he described wild rice, for instance, he conflated the usage of colonial and indigenous populations to a point where speaking of indigenous knowledges becomes impossible, instead hinting at broader mixtures of French and indigenous knowledges that ultimately remained unexplored in his scientific texts. He wrote simply that "our *voyageurs* who go into the *pays d'en haut* eat it with pleasure, they cook it as well as the *sauvages* with the fat of meat."[124]

At least some of this may have been a reluctance on the part of aboriginal peoples to share their information with colonial naturalists with whom they had had little contact and with whom there was little prospect for a long-term relationship. Indeed, while the work of historians such as Gilles Havard and Allan Greer has hinted that the quotidian exchanges between colonial and indigenous populations at the margins of empire may have produced new forms of ecological and botanical knowledge, there is little evidence to suggest that colonial naturalists looked for, or were exposed to, the knowledges of aboriginal peoples.[125] The only inclusion of aboriginal knowledge in the work of Alexandre Vielle, for instance, comes from one of "several persons who have had *sauvages* from Carolina as their slaves."[126] Providing information on the *cirier* (wax myrtle) that had been requested of him by the Marine, it seems equally possible that here the presence of aboriginal informants served as a visible claim to an epistemologically privileged geographical position rather than as evidence of an interest in or respect for indigenous ecological knowledge. Additionally, as in this case indigenous knowledge provided little beyond a confirmation of his own empirical findings, Vielle seemed reluctant to grant legitimacy to any knowledge acquired through his own encounters with indigenous peoples. The result was that aboriginal peoples were almost altogether absent from the scientific accounts of North American flora produced on either side of the Atlantic, sharing a silence also frequently imposed upon colonial populations.[127]

*   *   *

By the eighteenth century, neither colonists nor French administrators believed that the environments of New France shared an essential familiarity with those in France. The practice of earlier generations of colonists, explorers, missionaries, and merchants had focused careful attention on features of the landscape and individual plants that strongly resembled French counterparts. The effect was that even where texts highlighted the differences associated with *sauvages* plants and places, New France's future as a colony presaged an era where these distinctions would dissolve under the careful hand of French cultivators. After a century of experience, new methods sought to understand colonial environments. Botanists, perhaps most familiar to us today, used the abstracting tools of their science to describe and catalogue novel species. Colonists emphasized the personal experience acquired from living in what was undeniably a new world to them, but as they recognized the limits of the knowledge systems that they had brought with them, they made new room for indigenous names, taxonomies, and ecological knowledge. Both were studies of difference and subtle recognition that New France was in fact a new place and not simply *sauvage* and in need of cultivation.

French Atlantic science was the product of ongoing tension between the need for the participation of colonists and indigenous peoples and the effort of the Académie to place itself at the center of new scientific networks that transformed Paris into a scientific capital. Bound to their colonial correspondents by their need for specimens, French naturalists such as Tournefort, Vaillant, and Bernard and Antoine de Jussieu were nonetheless able to leverage their privileged access to the legitimacy and resources of the Académie to recruit and discipline royal physicians such as Sarrazin, Gaultier, Prat, and de Fontenette and apothecaries such as Alexandre Vielle. Establishing the scientific and social conventions of Parisian science in networks of Atlantic and continental communication, both Parisian naturalists and colonial correspondents helped fashion Paris into a scientific capital and reduce France's North American colonies to scientific peripheries.

The story of correspondents such as Sarrazin also reveals, however, that the French Atlantic was a far more crowded place than the records of the Académie Royale des Sciences and Jardin du Roi would have us believe. As previous chapters in this book have shown, the records of the Jesuits and French merchants who also took an interest in colonial flora demonstrate clearly that the networks the Académie extended into the North American continent were not alone in circulating novel plants and in maintaining networks of collectors and informants. Colonial naturalists such as Michel Sarrazin were only able to succeed in fashioning themselves as authoritative voices on North American flora because of their ability to co-opt existing conversations about colonial environments between colonists and indigenous peoples. Sarrazin and his colleagues became mediators between the Atlantic networks of the Académie that were maintained by the affective bonds of patronage and the profit-centered networks of amateur collectors that stretched into the interior of the American continent. Extending the reach of the Académie and establishing themselves as key nodes in Atlantic networks, these colonial correspondents were simultaneously implicated in the delegitimation of the activities and knowledge of colonial and indigenous plant collectors.

CHAPTER 6

How New Was New France?

On Saturday, March 12, 1718, the Jesuit Joseph-François Lafitau presented a "little book" to the members of the Académie Royale des Sciences who had gathered for its twice-weekly meeting in Paris. This book outlined his controversial claim that he had discovered ginseng (*Panax quinquefolius*) in the forests south of Montréal near the mission at Kahnawake at which he had lived and worked.[1] The minutes were virtually silent, however, on the content of the book that Lafitau delivered, aside from adding a note that it was "sur le Gin-seng."[2] It appears that the members of the Académie took little notice. There is no mention of any accompanying speech or introduction to the text, and no response to the Jesuit's claim to have discovered ginseng was recorded. But even if this was the Jesuit's first visit to the Académie in person, it was the third time in fifteen months that the Académie had discussed Lafitau's investigations of ginseng. Over the course of several years, members of the Académie such as the botanists Antoine-Tristan Danty d'Isnard, Antoine de Jussieu, and Sébastien Vaillant debated the identity of the plant that Lafitau had discovered in New France and, if it was in fact ginseng, who deserved the credit for finding it.

Part of the controversy that emerged around Lafitau was a response to his efforts to foreground the participation of indigenous peoples in his discovery. In the account of his search for the plant that was published in 1718, Lafitau recounted the three months during which he had searched for the plant outside of Montréal and his mission at Kahnawake, as well as the moment when he first suspected that his labors had borne fruit. "Having eagerly pulled it out [of the ground]," he wrote, "I carried it joyfully to a *Sauvagesse* who I had employed to search for it."[3] He listened carefully to this Mohawk woman's botanical identification and provided his readers with information about the use of the plant in her culture. In his efforts to publicize and defend his dis-

covery, Lafitau staked much of what he claimed to know upon his close interaction with indigenous cultures of North America. He privileged the role of
the Society of Jesus as intermediaries in the position to speak for both exotic
flora and the indigenous cultures who knew its secrets. "I report," he explained,
"only what I have learned from my *Sauvages*."[4]

Lafitau did not claim, as Champlain had done a century before, that
North America offered a distorted version of France that could be cultivated
and made recognizable. An entire century of encounter with new places and
peoples had sufficiently demonstrated the limits of French knowledge and the
ability of French horticultural practices to efface observable differences. As
Lafitau reached out to aboriginal communities for help in understanding
American flora, he reflected a broader trend among colonists and other missionaries who similarly sought insights from indigenous knowledge. In this
he differed substantially from earlier generations of colonial authors who dismissed and marginalized indigenous knowledge not only as deficient but as
the cause of the differences between European and American flora. Yet Lafitau
nonetheless remained committed to substantial facets of the ideology of cultivation that had driven much of the first century of French colonization in
North America. Lafitau insisted on the existence of biological connections between Europe and a New World that, observed properly, was not so new at all.
Ginseng offered the Jesuit an opportunity to resist the encroachment of a new
botanical science that highlighted the novelty and fundamental difference of
species such as American ginseng. If Lafitau no longer sought to cultivate a New
France, he nonetheless continued to believe that the environments of French
North America offered evidence of global continuities and connections, botanical augurs that testified to the colonial future of northeastern North
America, its environments, and its indigenous peoples.

As American ginseng became the focal point of competing narratives of
discovery, the fate of the plant itself—how it was identified, named, and
understood—therefore became inextricably tied to a larger debate about how
to know New France. The members of the Académie who studied the plant
sought to center the production of knowledge in Paris and to privilege the scientific networks of the Atlantic world. Akin to what Ralph Bauer has called
the "epistemic mercantilism" of the Spanish and Anglo-Atlantic empires, the
Académie sought to map a division of scientific labor onto the French Atlantic world.[5] While influenced by the methods common to contemporary botanical sciences, Lafitau's study of ginseng ennobled the otherwise unremarkable
plant by associating it with the global mission of the Society of Jesus and

mobilizing it as evidence for human and environmental continuities between the Old World and the New. For naturalists such as Vaillant, Jussieu, and Danty d'Isnard, ginseng was instead defined by its floral structure, and the analysis of each pistil and stamen inserted American ginseng into a scientific economy where the plant was principally valued for its novelty as a new species. The debate over the identity of ginseng therefore became an avenue to dispute who could claim to know New World environments and the methods through which knowledge was made. The question of whether American ginseng was a new species or not became an opportunity to debate how new New France was.

*   *   *

Scientific authorities in Paris had begun to hear about the possibility that ginseng had been discovered in North America in 1717 as both Lafitau and Michel Sarrazin, colonial administrator, royal physician, and elected correspondent of the Académie Royale des Sciences, sent descriptions of the plant and accounts of its discovery. Writing to the Abbé Jean-Paul Bignon, then president of the Académie, Sarrazin apologized for failing to discover the identity of a plant that had been hiding in plain sight in France's North American colonies, attributing his failure to his lack of exposure to indigenous cultures and languages.[6] The frustration and disappointment that he felt was palpable when he wrote to the Académie that "there appears here a plant that is believed to be the geinseng of Tartarie or of China, that the *sauvages* have found and that they have given to the Jesuits: they have made their accounts and we rest in the dark . . . I have been a botanist here for twenty years and yet this plant has unfortunately escaped me."[7] Yet he still assured the Abbé and Académie that "I am sending live roots of geinseng to the Jardin du Roi."[8] The letter thus reads like a botanical mea culpa, evidence that France's first royal botanist in its North American colonies felt that he had failed his patrons and his Parisian colleagues.

Sarrazin had first collected the plant he now knew to be in ginseng in 1700, when he sent specimens of what he had called *Plantula Marilandica* to Joseph Pitton de Tournefort. His study of the plant is preserved in a manuscript written by Sébastien Vaillant (Tournefort's successor at the Jardin du Roi) but based upon Sarrazin's own descriptions. Sarrazin's method was entirely superficial and designed to record the plant's unique morphological character that alone could provide an adequate identification. "Its root," he wrote, "is

fleshy like that of the *Ornithogalum*." Sarrazin continued, as he explained
that "it produces a single stem so delicate that it is impossible not to break it as
it is collected. It is about a foot long and produces two or three leaves arranged
in a collar supported on inch-long stalks."[9] Only his own observations were
featured, and while other entries contained information about medical or
economic uses of Canadian flora, here there was none.

When he wrote in 1717, Sarrazin was clearly anticipating the arrival of
news of Lafitau's discovery of ginseng in France and seemed ready to admit
the superiority of the Jesuit's methods. In fact, word of Lafitau's discovery first
appeared in print in the January 1717 edition of the *Journal de Trévoux* titled
"Le Genseng, Plante si precieuse à la Chine, découverte dans le Canada." The
Journal de Trévoux was a predictable venue for an article that, while authored
by Jesuits, was aimed at an audience beyond members of the Society of
Jesus.[10] Jesuits had become an influential mediator in the transmission of indigenous plants and ecological knowledge within the colonial Americas and
across the Atlantic. By the time that French Jesuits had arrived in North America in 1611, the Society of Jesus had contributed substantially to the natural
history of the Americas.[11] The emphasis placed on the senses in the *Spiritual
Exercises* and the sanctification of study and education encouraged a commitment to empirical observation and rational study of the natural world.[12] From
missions scattered throughout North America, the many authors who contributed to the annual *Relations* and the later *Lettres édifiantes et curieuses* provided their European audience with unique insights into regions of the continent
rarely visited by European observers capable of describing them.[13] This account was anonymous, but based on similarities between some of the phrasing in the *Journal de Trévoux* and Lafitau's 1718 *Mémoire*, it is clear that even
if Lafitau was not the sole author of the article, the text was at least based on
some of his written work.[14]

The narrative of discovery published in 1717 traced out a global Jesuit communication network that stretched from Asia to Kahnawake. The account
explained that Lafitau had been inspired by a letter that a confrère had written to the "Procureur-Général des missions des Indes et de la Chine" from
Beijing on April 12, 1711, and that had been distributed globally as part of the
tenth edition of the collection of missionary reports known as the *Lettres
édifiantes et curieuses* published in 1713.[15] The *Journal de Trévoux* did not specifically name Pierre Jartoux as the Jesuit who had inspired Lafitau's research.
Yet it seems clear that Lafitau's work was presented as an extension of Jartoux's
pioneering botanical work in East Asia. Jartoux, a Jesuit who had been in

China since 1701, had seen the plant in the region that he called "Tartarie." As a participant in a mapping project undertaken by the Society of Jesus for the Kangxi emperor, he saw the plant collected and was provided with information on its medical usage by fellow travelers.[16] His account was the first to provide a complete botanical description of ginseng as it existed before being processed and readied for trade as a commodity.[17] While Asian ginseng had been arriving in Europe since the expansion of the Dutch trading presence in East Asia in the early seventeenth century, the botanical identity of the plant had long remained elusive. The specimens that arrived in Europe as shriveled roots had been prepared for a market far larger than the handful of European botanists interested in plant morphology and lacked features such as leaves, flowers, or fruit necessary for botanists to scientifically identify them.[18] Jartoux's account therefore found an eager audience in Europe and was soon endorsed by the Académie Royale des Sciences.[19] For Lafitau, Jartoux's text both offered descriptions of Chinese usage and Asian ecosystems and provided a tantalizing suggestion: based on a preliminary comparison of the mountainous landscapes that he had experienced in Asia and those that he had read about in North America, Jartoux speculated that ginseng might also be found in the place that he called Canada.[20]

By 1717, Lafitau had spent five years at the mission at Kahnawake and had acquired a respectable knowledge of local languages and cultures that he foregrounded in the account of his discovery.[21] Readers of the *Journal de Trévoux* were cautioned that Lafitau's experience with the plant was limited, but Lafitau wrote to assure them that "the *sauvages* most knowledgeable of simples (medicinal plants) have similarly assured this missionary that they use it with success for several sicknesses."[22] Lafitau focused particular attention on developing an etymological analysis of the Chinese and Iroquoian words for the plant. "This plant is called *garentoguen* in Canada, & this Iroquoian word signifies the same thing as the Chinese word ginseng," the article argued.[23] Both meant roughly "representation of man, [or] resemblance to man."[24] It was this cultural knowledge, and the comparisons with the Chinese use of ginseng that it permitted, that Lafitau presented as his principal evidence of the American plant's identity.

Providing evidence not only that the plant existed in both American and Asian ecosystems, the article in the *Journal de Trévoux* hinted at an argument that would later define Lafitau's study of American aboriginal cultures. Continuing, this article conjectured that "two names so similar in their signification could not be given to the same thing without a communication of ideas, &

by consequence of people: from this one could conclude that these oriental *Tartares* whose customs resemble those of the *sauvages*, are not so distant from Canada as one thinks."[25] Before he turned his attention to a comparative history of non-Christian peoples the world over, Lafitau explored a previously obscure American plant as a route toward producing a global history. His discovery of ginseng was an opportunity to instruct readers about the humanity of these indigenous peoples and to explain the value that they and their knowledge could bring to France.

As the reports of both Lafitau and Sarrazin crossed the Atlantic, and even as they represented divergent academic traditions and approaches to the indigenous cultures of New France, they now shared a common appreciation for the role that aboriginal peoples and their knowledge of American environments could play for colonial investigators. While later disputes divided authors more simply into academic and Jesuit, these first whispers of a remarkable discovery in the forests outside of Montréal belie any such distinction, pointing instead to emergent—if still heterogeneous—colonial knowledge that had not categorically ruled out the participation of Native peoples. The knowledge that they sought to circulate was the product of long-term and multidirectional exchanges with their indigenous neighbors.

*   *   *

Although the article that appeared in the *Journal de Trévoux* in January 1717 appeared to offer the first written presentation of Lafitau's discovery, it also suggested that at least some contemporary naturalists at the Académie had already responded critically to what the Jesuit proposed. Even as he dedicated his *Mémoire* to the regent Philippe d'Orléans, Lafitau was clearly appealing directly to the Académie Royale des Sciences to legitimate his work. In submitting his work for approval to this body and its members, Lafitau therefore took part in a broader centralization of the Atlantic networks in which knowledge was produced, circulated, and authorized the central role of the Académie.

While Jesuit study of ginseng in North America and Asia had focused as much on the human as the natural history of the plant, in Paris their accounts were judged by their botanical descriptions alone.[26] The Académie first discussed the Jesuit's work at the Académie's meeting of January 16, 1717. While the *Journal de Trévoux* neglected to name Lafitau's early detractors, the physician and botanist Danty d'Isnard was soon Lafitau's most public critic in

Paris. A member of the Académie since January 1716, Danty d'Isnard was also a former botanical demonstrator at the Jardin du Roi.[27] By the time the Académie met on the fourth of December of that same year, Danty d'Isnard had prepared remarks "on the ginseng of China and that of Canada" that revealed both his lack of faith in the Jesuit method of studying Chinese and North American floras and a demonstration of the means that he considered appropriate in their place.[28] Specifically, the academician turned to a Dutch physician named Engelbert Kaempfer as an authority on the ginseng found in Asia. Kaempfer had traveled to Japan via Russia and the Middle East and had established himself as a reputable source of information about the extra-European world as he worked for the Swedish crown and, later, the Dutch East India Company. His description of ginseng, produced "under the botanist's knife," he wrote, exhibited the same focus on plant morphology that defined the work of contemporary European botany.[29] Ginseng, or *ninjin* as Kaempfer called it, appeared as part of his *Plantae Japonicae*, a text that he wrote about his botanical research between 1689 and 1692 when he had resided in Nagasaki.[30]

As a physician himself, it was possible that professional chauvinism inclined Danty d'Isnard to favor Kaempfer's description. In December 1717, Danty d'Isnard attempted to refute statements about Lafitau's discovery line by line. He quickly declared his support for Kaempfer's research and systematically repudiated the Jesuits' claims that the German physician's work had been inferior to their own. Whereas the Jesuit authors at the *Journal de Trévoux* had asserted, for instance, that "Mr. Kaempfer says, that the plant that he is describing, grows in China, it does not grow there, the Chinese themselves assure it," Danty d'Isnard countered that by citing "the statements of Mr. Kaempfer that one obtains Ginseng from Korea, and that it grows on the mountains of this country, and not in China."[31] While the Jesuits had been magnanimous enough to suggest that Kaempfer may have actually described Asian ginseng even if he was mistaken in some key details, Danty d'Isnard rejected both Jartoux's and Lafitau's accounts equally.[32] As he critiqued Lafitau, he also provided concrete clues about why he preferred Kaempfer's descriptions. In addressing his concerns to the gathered members of the Académie, Danty d'Isnard highlighted the facile equivalencies established between Asian and American plants and the descriptions provided by Kaempfer and Jartoux. After quoting the Jesuit statement that "after all the difference is not so great between the description of P. Jartoux, and that of Mr. Kaempfer, that they could not fit the same plant" that had appeared in the *Journal de Trévoux*,

Danty d'Isnard carefully demonstrated that the Jesuit account was rooted in observational error.[33] The critique of the Jesuits' use of ethnographic data is clearest in his refusal even to discuss it. He ignored, for example, the analysis and comparison of the plant's medical usage in Asia and North America that Jesuits had forwarded as evidence of botanical continuities between East Asia and northeastern North America. At the bottom of a page littered with scientific names of morphological features, his description finished simply that "one only need read these two descriptions and see these two figures of very different plants."[34]

Lafitau found at least some support among the members of the Académie. While Danty d'Isnard hinted that he knew other members of the Académie who shared his concerns about Lafitau's claims, the botanist Sébastien Vaillant soon publicly stated his belief that ginseng had been found in North America. Yet Vaillant's support soon revealed itself to be as problematic for Lafitau as Danty d'Isnard's criticism had been.[35] On February 9, 1718, only two months after he had first publicly criticized Lafitau's research, Danty d'Isnard read a paper written by Vaillant (who did not attend the meeting himself) to the members of the Académie. The paper established a new genus of plant named Araliastrum and was published later that same year as an appendix to Vaillant's *Discours sur la structure des fleurs, leurs différences et l'usage de leurs parties*.[36] Vaillant's text identified two other genera as well, with all three serving as models for a new means of identifying plants that focused on floral structure and that, prefiguring the contributions of Carolus Linnaeus, suggested that plants reproduced sexually.[37]

Vaillant's text announced the creation of "a new genus of plant named ARALIASTRUM of which the famous Ninzin or Ginseng of the Chinese is a species."[38] Like Joseph Pitton de Tournefort before him and in following the standard practices of European botany since the early seventeenth century, Vaillant focused his description on the botanical genus rather than the individual species.[39] This meant that individual species could be distinguished with the greatest possible economy, highlighting only the features that differed from the more abstract description of the genus.[40] Describing this genus, he wrote that "the Araliastrum is a genus, . . . of which the flower . . . is complete, . . . regular, polypetaled, & hermaphroditic, . . . the ovary crowns the calyx which has several points, becoming a base . . . , in which are ordinarily found two flattened seeds, cut like a kidney or in half circle, which represent together a type of heart. Additionally, the simple stem is terminated by an umbel, of which each pedicel carries only one flower, & this root is attached above its middle, like that of the Anemone, by a circular assemblage of some stalks, at the end of which appear several leaves disposed in rows."[41] Ginseng was further identified below as the first of four species of Araliastrum and given the name *Araliastrum Quinquefolii folio, majus, Ninzin vocatum, D. Sarrazin*. In a text where morphological descriptions drowned out discussions of medicinal properties or commercial value, ginseng was therefore only recorded as a species of Araliastrum that was differentiated by its five large leaves.[42] There was no mention of Lafitau, the Haudenosaunee, or even North America in this description. The only other information attached referenced the Académie's sole North American member, the royal physician of New France Michel Sarrazin, and the Japanese name for the plant, *Ninzin*, that had been used in Kaempfer's published work. Vaillant's account inserted the plant into networks in which authority circulated among a small body of trained and reputable botanical observers and published botanical texts.

Vaillant began his own discovery narrative earlier than the contributions of either Jartoux or Lafitau. He explained to his audience that the species of Araliastrum "are common in Canada, from where Mr. Sarrazin councilor to the Superior Council, Royal Physician and Correspondent of the l'Académie Royale des Sciences had sent [them] to the Jardin du Roi of Paris for the first time in the year 1700."[43] In 1708, Sébastien Vaillant compiled the letters, plant lists, and specimens sent from Canada by Sarrazin and produced the first French botanical text devoted exclusively to North American plants, the *Histoire des plantes de Canada*. This manuscript reaffirmed the name *Plantula marilandica*, which had been attached to specimens of the plant sent from Canada in 1704, and stated that Sarrazin had sent the plant to Paris as specimen no. 73 in 1700.[44] Even if it was clear that neither Sarrazin nor Vaillant knew that *Plantula marilandica* was ginseng in 1700, 1704, or 1708, the narrative that Vaillant produced in his 1718 *Discours* privileged the preliminary investigations of the plant's morphology and erased the subsequent contributions of the Haudenosaunee and the Society of Jesus beyond their role in providing the Académie with specimens. In fact, Vaillant's narrative of the American discovery of ginseng was a sort of non-narrative, demonstrating clearly that while plants could be found in North America, they could be studied and named only in Paris. Vaillant implicitly denied that the plant had had a history before it was studied by members of the Académie. In privileging Sarrazin's earlier efforts to transport American flora as they studied ginseng, Vaillant and the Académie made a powerful statement to collectors such as the Jesuits who sought to bypass the Atlantic networks of the Académie and produce their own botanical research: true discovery could happen only in Paris.

* * *

At eighty-eight pages, Lafitau's 1718 *Mémoire* extended rather than transformed the argument first presented in the *Journal de Trévoux* in January 1717. Lafitau used these additional pages to prove that both he and Jartoux had seen, touched, tasted, and described the same plant, and he layered detail upon detail, rhetorically demonstrating his own authority as witness and guiding his readers to discover the plant for themselves through his account. The text focused on developing a rich first-person narrative of Lafitau's encounter with the plant that walked the reader through the forests of New France and the mission of Kahnawake before turning to train his reader to see the plant as he had through an analysis of its morphology that proceeded inch by inch and from root to flower.[45] The *Mémoire* therefore sought to prove the legitimacy of Lafitau's methods by re-creating his experience for his readers as well as by drawing upon supporting evidence from the global networks of the Society of Jesus.

Lafitau explained that the business of the mission at Kahnawake had brought him to Québec in the fall of 1715, where he took the opportunity to read the "collection of edifying letters of the missionaries of our company who work throughout the world."[46] This was neither new nor out of the ordinary, as reading the collected letters of his confrères was, he wrote, "a powerful motive to sustain the difficult work of our missions with constancy. In effect nothing is more capable of softening our struggles, & of animating us, than the example of those of our fathers who are in the same situation as us, who appear to disregard their fatigue, and consider themselves happy when it has pleased the Lord to give some success to the Gospel that they preach, or to console those for whom obstacles & setbacks render their work sterile."[47] Lafitau conceived of the Jesuit enterprise in global terms.[48] Reading the accounts written by other Jesuits working throughout East Asia, Lafitau was inspired to continue to expand the scope and scale of his own work in the Saint Lawrence Valley, not simply acting within a global framework but giving life to communication networks that stretched from one end of the earth to the other.

In such a manner, Lafitau became both an individual observer and a node in a global network. Yet as his *Mémoire* slipped between travel narrative, scientific treatise, and ethnographic study, Lafitau also made his debt to indigenous peoples explicit. He was unimpressed with what nearby colonists and other missionaries had thus far learned from indigenous cultures and lamented that the Jesuit missionaries in North America had missed opportunities for acquiring indigenous botanical knowledge that had been profitable for the Society of Jesus in other missions and on other continents; he singled out Peru and Brazil as examples of where Jesuit investigations into indigenous botanical knowledge had brought new drugs to Europe and praise and profit to the Society of Jesus.[49] Lafitau wrote that he had turned to local Mohawk women with whom he lived and to whom he ministered for help finding the plant. Yet he was only able to produce an approximation of the image provided in Jartoux's account from memory and his aboriginal informants were of little help at this early stage. He confirmed Jartoux's hypothesis only with the arrival of summer and the appearance of the plant's vermillion fruit that had been described in such detail in the *Lettres édifiantes*.[50]

Lafitau's initial observations of the ginseng that grew near Kahnawake suggest considerable similarity between contemporary and eighteenth-century populations of a "widespread but scarce understory plant."[51] He made his discovery in deciduous woods that provide considerable shade and low evaporation.[52] His *Mémoire* provided a description of the plant growing in isolated clusters throughout the "diverse cantons where they grow the one next to the others."[53] They were composed, he explained, of plants that had grown 3 or 4 leaf clusters, plants that are typically 7–9 (3 leaf clusters) or 10–11 (4 leaf clusters) years old.[54] The number of plants that he described in each location suggests that the sites would not support extensive harvesting. When he wrote, for example, that "never more than seven or eight roots are found" in the same place, he provided numbers substantially below those found in protected populations in recent decades.[55]

Lafitau expressed doubts about the sustainability of widespread collection that are borne out by recent ecological research.[56] At the northern reaches of its natural range, it is likely that colonists and indigenous people who, Lafitau wrote, would have had "to search further in the woods" would not have found any further north. The clearance of land for agricultural use could have removed some of the plant's habitat. Likewise, the increased disturbance of forests for activities such as maple syrup collection would have removed at least some habitat and limited the plant's reproductive ability.[57] It is also likely that some natural predation from deer would have decreased with the increase of human populations and the diversification of the fur trade.[58] If the collection of the plant by Lafitau and Sarrazin would have had a negligible effect on local populations, ginseng was a plant that could little afford its fifteen minutes of fame. Any more collection would threaten the ability of Native Americans to acquire "one of their ordinary remedies" in the forests near Montréal.[59]

Lafitau maintained the complex and ambivalent relationship with indigenous knowledge already in evidence in the first article about his discovery, and he alternated between crediting himself with the intellectual heavy lifting and revealing his dependence on local networks of knowledgeable women. He wrote, for example, that "the questions that I had asked the *Sauvages* about Gin-seng did not advance me very much."[60] He argued for the self-evidence of his discovery, claiming that "I did not consider it for long without suspecting that this could be the plant for which I was searching" and positioning himself as a credible observer.[61] Yet he felt the need to authorize other local observers as well, noting that he had immediately gone to an indigenous woman to confirm his discovery: "having eagerly dug it up, I carried it joyfully to a *Sauvagesse* who I had employed to search for it. She recognized it at once for one of their ordinary remedies, and told me the types of usage that the *Sauvage* had for it," he explained.[62] From nearby aboriginal peoples, Lafitau learned that the plant was used as a purgative and from Wendat and Abenaki informants he learned that it was used to treat dysentery.[63] To these observations he added his own, testing the plant's qualities himself. "These responses and the experiment of the *sauvagesse* of which I have already spoken, who was cured three times of fever," wrote Lafitau, "were all that I knew when I sent the Gin-seng of Canada to Paris and that the Père le Blanc had the honor to present to your royal highness. I had experimented on myself, and I was persuaded that by its use I was cured of a bout of rheumatism of which I was very tired, and which I have not felt again. I have since used some for a flux of blood that I beat with a single dose."[64] Authorial voice and authority were therefore not always synonymous as, where Lafitau sought to broaden the sources used to identify the plant, he included a cast of indigenous actors and methods that diverged dramatically from those relied upon by members of the Académie.

Lafitau's success was based upon a willingness to work with Haudenosaunee women, and it was while working in close proximity with them that he became aware of the etymological association between the Iroquoian *garentoguen* and the Chinese ginseng. His "surprise was extreme" when he learned, via Jartoux and the Jesuit polymath Athanasius Kircher, that the Chinese had named the plant for its "*Resemblance of man*" or to "*Man's Thighs*."[65] Lafitau argued that these were all simply different words for the same object. "In effect *Garent.oguen* is a word composed of *Orenta*, which signifies the thighs and the legs, & d'*Oguen*, which means two things separated," he explained.[66] Ginseng could be discursively dissected to produce the same etymology.[67] For Lafitau, this was the crux of his discovery and revealed that he was searching not only for an Asian plant but for the roots of a unified human and natural history of the world. The narrative of his discovery suggests that he grasped the significance of this etymological evidence immediately. "I could not stop myself from concluding that the same signification could not have been applied to the Chinese word & to the Iroquoian word without a communication of ideas, & by consequence of people," he wrote.[68] He then followed this evidence to a conclusion that would have seemed remarkable to any audience that simply expected a description of American ginseng. "By this I confirmed an opinion that I already had," he wrote, "that America was the same continent as Asia, to which it was connected by Tartarie to the north of China."[69]

The engraved plate that Lafitau had produced (Figure 10) represented this entanglement of cultural and natural observation visually. When Sébastien Vaillant was working on identifying ginseng, he wrote that "between the parts that characterize the plants, those that are called flowers, are, without contradiction, the most essential."[70] Yet while representations of the plant's floral structure and leaf shape were present in Lafitau's image, it was the root itself that dominated the composition and that, by extension, defined the plant. The fixity of the leaf shape and floral structure was conveyed by their uniform representation and a singular close-up of a leaf. Lafitau, however, overwhelmed the composition of his image with representations of ginseng's root, a part too irregular to be of use for taxonomic purposes by French and European naturalists. His image recognized and even emphasized this irregularity as he visually suggested that the root was nonetheless the key to understanding the nature of the plant and to properly identifying it. This mirrored a textual description of the plant that started with, and lavished attention upon, the shape, texture, color, and taste of ginseng's roots. This focus on ginseng's roots was also present in the preparation of the physical specimens that Lafitau sent and brought to Paris.[71] Lafitau's focus on ginseng's roots quite literally turned French botanical norms on their head.

In writing the name of the plant across the top of the plate in his *Mémoire*, Lafitau also highlighted it as an essential facet of the plant itself, more than a convenient tag. Written as "Aureliana Canadensis, Sinensibus Gin-Seng, Iroquœis Garent-Oguen," it was the plant's name and the associated etymological and ethnobotanical evidence that provided key evidence of con-

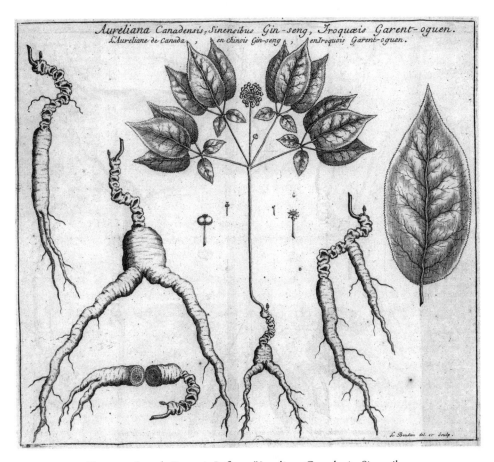

Figure 10. Joseph-François Lafitau, "Aureliana Canadenis, Sinensibus
Gin-seng, Iroquœis Garent-Oguen," *Mémoire . . . concernant la précieuse
plante du gin-seng*, 1718. Courtesy of the John Carter Brown Library
at Brown University, Providence, Rhode Island.

tinuity between Chinese and Iroquoian cultures and that justified his identification of the roots as the most important part of the plant.[72] Like the roots
in the image, the differences in the name were only apparent; both *garentoguen*
and ginseng demonstrated a common focus on the human-like shape of the
plant. Manifest differences masked an essential continuity that proved that
the morphology of the plant and its "idea" were inseparable. Lafitau argued
that, properly applied, botany could answer world historical questions and
refute the claim that the Americas were a New World.

Lafitau's study of ginseng was inseparable from his larger concern for the spiritual well-being of his Haudenosaunee charges. His visit to the Académie in 1718 was a break from a trip to Paris that was otherwise focused on securing permission to move Kahnawake to new, more fertile lands that would better support agriculture. He also worked to highlight the costs of the fur trade that brought liquor into indigenous communities, and he urged the French crown to crack down on the illicit trade.[73] The two subjects of Lafitau's research, American aboriginal cultures and ginseng, therefore alternated and competed to become the principal focus of the text. Yet this missionary-naturalist remained a missionary first and foremost and his primary goal remained the salvation of American aboriginal communities. For even as Lafitau couched his interest in ginseng in a broader concern for uncovering potentially useful aboriginal medical and ecological knowledge, ginseng was ultimately "a clinching piece of evidence for his argument that North America was joined to the Asian mainland somewhere north of China."[74] His desire to establish cultural continuities between Asia and North America informed his search for botanical continuity and was at the heart of his text. Lafitau again hinted at the influence of the global mission of the Society of Jesus on his project and, specifically, the work of other Jesuits in China who had argued that Chinese histories showed evidence of a Judeo-Christian origin. Although their defense of Chinese culture was not without controversy, Jesuit authors such as Jean-François Foucquet and Joachim Bouvet argued that Chinese texts that described the founder figure Fu Xi were actually acknowledging the Christian God.[75] They similarly pointed to specific Chinese "ideographs" as evidence for a single origin of both Chinese and European cultures.[76] As with ginseng and *garentoguen*, differences between European and foreign cultures were only skin deep, masking a more fundamental continuity that provided evidence of a shared human history.

Ecological and biological continuities were foregrounded as evidence of the cultural continuities that Lafitau claimed located American plants and peoples in a universal history. Comparisons of latitudes, climates, topography, and—of course—the morphology and ecology of a single American plant with its Asian counterpart flattened differences between Asia and North America to make them both knowable. The *Mémoire* was an effort to locate American indigenous peoples within a broader history of continental migrations as much as it was an effort to prove the Asian origins of an American plant. Both in his *Mémoire* and later work, Lafitau's challenge was nothing less than to "demonstrate rationally that the Amerindian difference was only, in effect, an

apparent difference, superficial and contingent, and to thus reduce Amerindian alterity to a familiar identity."[77] Questions surrounding the identity of
ginseng and that of aboriginal cultures were inseparable, and evidence of historical migration testified to both a terrestrial bridge between Asia and the
Americas and the common origins of all humanity in Adam and Eve.[78] Lafitau's text anticipated some of the major claims that would be made in his 1724
Moeurs des sauvages américains; both were clear arguments against theories of
multiple independent creations.[79]

If he no longer called for the cultivation of a New France, Lafitau nonetheless saw colonialism as an opportunity to examine and accentuate underlying affinities between Europe and North America. In his *Mémoire*, he took
the botanical knowledge of the Haudenosaunee seriously even as he subjected
it to the same sort of dissection that botanists normally reserved for plants. Yet
to acknowledge indigenous knowledge was to simultaneously acknowledge
its limits. "I must admit that we do not yet know this plant well enough," he
wrote, "because we know it only through the *sauvages*, the Chinese, and the
Japanese who are at base bad physicians, little instructed in the principles of
anatomy and the rules of the art."[80] He therefore suggested that the Haudenosaunee were largely unaware of the significance of their own botanical knowledge and he denied that they were capable of joining a scientific discussion in
their own right. Lafitau saw indigenous knowledge as a valuable shadow of an
original knowledge shared by all humanity, as he wrote that "necessity has
made the *Sauvages* doctors and herbalists; they search plants with curiosity,
and try all; . . . they have found [medicinal plants] by long usage which for
them takes the place of science."[81] Indigenous knowledge was incomplete and
could only be understood through a comparative analysis that highlighted
continuities and downplayed local particularities. In comparing the usage of
ginseng between American and Asian contexts, he sought to distill the true
properties of the plant and the most efficacious deployment of them. The next
step, he wrote, was to chemically analyze the roots.[82]

To argue for the legitimacy of indigenous knowledge, Lafitau theorized
that neither ginseng nor aboriginal peoples were originally indigenous to North
America. Ginseng, he wrote, was once also known in western Europe as mandrake. He hypothesized that, prized for its unique properties still evident
among the indigenous cultures of French North America and China, the plant
had been driven to extinction in Europe. Analyzing passages in the work of
Theophrastus that described the plants *ferule* and mandrake, Lafitau suggested
that the "superstitious" practices related to the collection and consumption

of the plant in the ancient Mediterranean could be compared with those that were still being practiced in North America.[83] Finding other plants in North America such as wild celery that had been associated with mandrake in the writing of other classical botanists such as Dioscorides, Lafitau in effect made an argument for a larger botanical migration from northeast Asia to prehistoric North America.[84] If Lafitau framed his efforts to bring the plant back to European gardens and pharmacopeias as a discovery, his *Mémoire* implied instead a reintroduction.

Ginseng was evidence that North America was another old world. There are echoes, therefore, of the project that encouraged early authors such as Champlain, Lescarbot, Sagard, and Biard to call for the cultivation of American plants and peoples. Yet Lafitau appreciated that material differences were far more intractable to French intervention. He did not propose that ginseng was in fact a French plant that had become *sauvage* through neglect. Nor did he suggest that the plant could be improved with French stewardship. Nonetheless, in his insistence that Asian and American ginseng were identical, he argued strongly against an emerging scientific consensus that focused on new, readily identifiable, and biologically distinct species. Lafitau's botany was, rather than the science of novelty performed at the Jardin du Roi and Académie, an affirmation of connection and a study of resemblance.

The commerce between New France and Old that Lafitau imagined was therefore as much in ideas as it was in roots and would enrich Native Americans and French people alike through a mutual rediscovery. Lafitau was unconcerned that the properties of the ginseng found in America might be different from that of Asia. After all, he wrote, plants were "almost everywhere the same."[85] He introduced botanical practices recorded by Jartoux and other Jesuit authors such as Jean-Baptiste du Halde, who had collected and studied Chinese medical knowledge from its foundational texts to prove this fact.[86] Where he noticed differences or lacked the basis for an adequate comparison of Asian and Iroquoian medical cultures, he experimented. For example, not content to simply find out the local uses of "one of their ordinary remedies," Lafitau encouraged the "*sauvagesse*" who had initially confirmed the identity of the plant to combat an intermittent fever with ginseng.[87] This was not a normal Iroquoian usage but instead represented an effort to confirm medicinal effects that had been observed in China and Europe. Believing Haudenosaunee knowledge to be universal in origin, Lafitau made it global as he sought to prove its legitimacy.

*   *   *

Even as Lafitau's work forced itself into the Académie's discussions of ginseng, the Académie's naturalists remained focused on highlighting the distinctiveness of American ginseng as a biologically unique species. As Danty d'Isnard and Vaillant had done before, two new narratives of ginseng's discovery produced in 1718 asserted the superiority of the Académie's scientific method and the centrality of its members in ever-expanding Atlantic scientific networks: Antoine de Jussieu's manuscript "Histoire du gin-sem et ses qualités" and the anonymous but thorough "Sur le Gin-seng" that was published in the 1718 edition of the *Histoire de l'Académie Royale des Sciences*. These two texts praised the contributions of the Society of Jesus to the study of ginseng but synthesized its research within accounts that privileged the role and authority of the Académie itself. The academic synthesis of 1718 demonstrated that the Académie acknowledged its dependence upon correspondents in the extra-European world but that it continued to define itself as the ultimate scientific authority in the French Atlantic world.

It is impossible to know exactly when the botanist Antoine de Jussieu wrote his "Histoire du gin-sem et ses qualités" although several textual clues suggest that it was written after the author had consulted Lafitau's *Mémoire*. It is also possible, however, that Jussieu had formed his narrative after consultation with Lafitau himself.[88] At the very least, Jussieu's manuscript demonstrated that its author had made himself well-informed about the botanical research of both Jartoux and Lafitau; both of their accounts featured prominently in his own. By the bottom of the first page of his manuscript, however, Jussieu instead highlighted the central role of the corresponding member of the Académie Michel Sarrazin, who, he wrote, had noticed ginseng "among the number of unique plants of that country."[89] Jussieu stated that Sarrazin deserved credit for the find because he had sent the plant, albeit named *Aralia humilis fructu majore*, to the Jardin du Roi in Paris by 1704.[90] When he recounted the evidence that Jartoux's descriptions of Asian ginseng matched those of the American plant, he visually prioritized and ranked the existing accounts on the page. First, he listed Sarrazin and the Englishman John Ray, both reputable botanists and the first to hint at ginseng's presence in North America. Lafitau was credited only with sending the plant to France.[91] When Jussieu's discussion turned next to the virtues of the plant, an area where he was again forced to rely on information collected by the Society of Jesus, he showed his hesitancy to rely solely on Jesuit sources. When he discussed the virtues of the leaves, for example, he noted that he had only Jesuit authors to rely upon. He wrote that these missionary-naturalists had claimed "that the leaves of this plant have a virtue approaching that of the root."[92] Yet Jesuit sources remained partial and could not match the "confirmed experience of the efficacy of the root" established by a multitude of Asian and European experience.[93] The contributions of the Society of Jesus were discursively marked as preliminary investigations that necessitated the confirmation of more reliable authorities such as himself and his Académie. Jesuit research into the etymology and cultural contexts of American and Asian ginseng went unremarked entirely.

The most thorough account of ginseng's discovery was published in 1718 as "Sur le gin-seng" in the 1718 edition of the *Histoire de l'Académie Royale des Sciences*. There is no record of when this *histoire* was presented to the Académie or who its author was. It is possible that it was also authored by Antoine de Jussieu and that the manuscript "Histoire du gin-sem" was an early version of the final printed text.[94] By 1718, however, other members of the Académie such as Claude Joseph Geoffroy were demonstrating their own interest in the virtues and identity of the plant.[95] Whoever it was, the author of "Sur le Gin-seng" was well read and had assembled all of the relevant texts that documented the Académie's history with the plant, expanding the narrative to include research that had never been mentioned previously. Predictably, the account began the narrative of ginseng's discovery not with Jesuit naturalists but with the introduction of the plant to the Académie in 1697. That year, the botanist and member of the Académie Claude Bourdelin read a report on the use of ginseng among the Chinese. The source of this information was not revealed, and this narrative therefore established the Académie itself as the original authority on the plant.[96]

Similar to Jussieu's manuscript, "Sur le Gin-seng" devoted a considerable amount of time to characterizing the contributions of Jartoux, Lafitau, and the Society of Jesus. Jartoux and Lafitau were both lauded for their work. Jartoux's description was praised as "the most exact & the best detailed that one had yet seen" when it appeared.[97] Yet this narrative also suggested that his principal accomplishment was the confirmation of what the Académie had already known or had, independently, begun to suspect following Bourdelin's investigations.[98] Unlike in Jussieu's manuscript, Lafitau received more attention than Jartoux in "Sur le *Gin-seng*." It even offered a backhanded affirmation of the legitimacy of the ecological knowledge of the Haudenosaunee, who were "very curious about plants, without being botanists."[99] This 1718 account was particularly appreciative of the level of detail in Lafitau's *Mémoire* that had been published that same year and of Lafitau's visit both to the Académie and to the Jardin du Roi. It acknowledged the early criticism of those such as Danty d'Isnard who had been more likely to trust Kaempfer than the Jesuit naturalists but explained that these doubts were laid to rest after Lafitau himself had come to Paris and explained his work in his *Mémoire*.[100]

Yet the praise lavished upon Lafitau was double-edged and echoed the accounts of Vaillant and Jussieu, who had sought to subordinate Lafitau's role in the narrative of ginseng's discovery to the Académie itself. Of Lafitau's *Mémoire*, it stated that "one sees there a description of the Gin-seng of *Canada* or *Garent-onguen* yet more nuanced than that of P. *Jartoux*." The author, however, soon diminished Lafitau's role. "One found in Lafitau's *Mémoire*," the author wrote, that ginseng's "virtues were proven by the P. *Lafitau* as much as he could at the time, & [were] the same as those of the *Mémoire* of M. *Bourdelin* & that common opinion attribute to Gin-seng."[101] So it was not that Lafitau or Jartoux was wrong, only that they were single observers who lacked the authority of the Académie. The dismissal of the missionary-naturalists closely mirrored the faint praise shown to the Haudenosaunee.[102] Lest the *histoire*'s audience overvalue the contributions of the Society of Jesus, it reminded its readers that the Académie had already been studying the plant for over a decade by the time the Jesuit research hinted at its true identity. "We were familiar with the plant before knowing that it was Gin-seng," it argued.[103] It went on to clarify that the rights of discovery went to the Académie itself as "M. *Sarrazin*, councilor and royal physician in Québec, very skilled botanist & correspondent of the Académie, was no sooner in Canada than he remarked it [ginseng] among the many unique plants of that country, and he put it under the name *Aralia humilis fructu majore* amongst those that he sent to M. Fagon in 1704 for the Jardin du Roi."[104] So even if the account reminded its readers that the arrival of ginseng and the subsequent enrichment of European medicine "[are] due as much as Cinchona to the Jesuit missionaries," it still restricted the role of the Society of Jesus in the scientific networks of the French Atlantic—networks that remained ever centered on Paris and the Académie Royale des Sciences.[105]

Sarrazin's role in the histories of the plant produced by members of the Académie only grew. This was most evident in a manuscript titled *Description de plusieurs plantes du Canada* written in 1749. The author of the text, Jean-François Gaultier, was also a corresponding member of the Académie and

Sarrazin's successor to the post of royal physician in New France.[106] Gaultier's description of ginseng was fairly long considering it was just one entry in his text, and it was detailed enough to demonstrate that the Académie's synthesis produced in the immediate wake of Lafitau's *Mémoire* not only survived but flourished in the Académie's Atlantic networks. Like Vaillant and Jussieu, respectively, Gaultier excused Sarrazin and Vaillant equally for failing to recognize that the plant they had named "araliastrum quinquefolii folio major" was ginseng.[107] Unlike these other authors, however, Gaultier also rehabilitated Sarrazin as a researcher and botanist in his own right. Gaultier, for example, asserted that it was Sarrazin who first introduced ginseng to the Jesuits as he grew the plant in their gardens for future study.[108] Gaultier also suggested that it was not Vaillant but Sarrazin who first hinted that it might be ginseng but that he lacked the proof to offer any conclusions with certainty.[109] Sarrazin's hardworking and cautious nature was repeatedly emphasized and was set in distinction to that of Lafitau, who was presented as both "fortunate [*heureux*]" and "more self-assured [*hardi*] than M. Sarrazin."[110] The choice of these words was significant as both indicated that Lafitau profited as much from luck and other people's research as he did from his own investigations. Calling Lafitau *hardi* was particularly telling as it subtly played with the multiple meanings of the word as courageous or self-assured and immodest or audacious.[111] While still recognizing the importance of Lafitau's contributions to the discovery of ginseng, Gaultier nonetheless signaled that Lafitau had not known his place. Sarrazin could be forgiven for not knowing the proper name of the plant because, far from Paris in Québec and Montréal, he could only ever produce a partial account of the plant. He had done what was expected of him when he sent specimens of the plant to Paris.

*   *   *

Lafitau's research spread word of the plant and its potential value before it had even been presented to the Académie Royale des Sciences in 1718. He described, for instance, the effect that his discovery had had on aboriginal usage of the plant in the Great Lakes region following the efforts of Louis de la Porte de Louvigny, a western post commander and military official in New France, who had encouraged its use.[112] Yet Lafitau was also anxious about the circulation of this knowledge to those who might be less interested in cultural commerce with indigenous peoples than simple trade. The Jesuit's encouragement to the French state to act quickly on his discovery was due in part also to a fear that if they did not, it was a virtual certainty that Protestant traders to the south would beat them to it. "The Flemish of New York will profit from it," he warned his readers, adding that "some of them have seen it sold at Montréal by the *Sauvages*, and it will no doubt be sent to England after this year."[113] Lafitau was aware that word of his discovery would spread, and he knew that while his own work proved that the plant had deep roots in Haudenosaunee culture, it likely also grew elsewhere in North America.

The early development of a commercial ginseng is, as one historian has written, "murky," and the geographical scale of the trade is difficult to state with any certainty.[114] Haudenosaunee and other indigenous peoples began supplying the plant to colonial merchants in the Saint Lawrence Valley by the 1730s. Residents of Kahnawake looked to the emerging market as an opportunity to supplement their income and diversify the local economy that had been undermined by both the western expansion of the fur trade within New France and declining populations of fur-bearing animals closer to home.[115] Over the seventeenth and eighteenth centuries a trade in "wampum belts, canoes, paddles, snowshoes, moccasins, [and] various craftworks which they sold in the towns on market days" had complemented a continuing commerce in furs.[116] In the eighteenth century both ginseng and Canadian *capillaire* (*Adiantum pedatum* or northern maidenhair fern) were harvested by Algonquian and Iroquoian peoples for export.[117] They became the most visible Canadian products in French pharmacopeias.[118]

Indigenous ginseng collectors were aided by merchants such as the Desauniers sisters, who became influential traders and smugglers of a variety of goods after acquiring land at Kahnawake in 1727.[119] The Desauniers sisters soon became an important and illicit connection to the markets of British North America via Albany and were active in the ginseng trade.[120] It was a mutually beneficial commerce. These merchants supported indigenous communities such as the Haudenosaunee who lived in between French and British North America and who resisted the erection of an imperial border through their territories. In return, the sisters and others like them were supported by their adopted communities when administrators targeted them for their role in the contraband trade in the 1730s and 1740s.[121]

Word of the plant and its value soon spread on both sides of the border and throughout the Atlantic world. Yet even if the intercolonial trade was widely embraced, its illicit nature makes tracing the circulation of either the root or information about it difficult.[122] Lafitau's confrère Pierre-François-Xavier de Charlevoix sought the plant out and found it among the Miami, who, he had discovered, also used the plant medicinally. "The *Sauvages* who apply themselves at all times more than others to medicine, make a great deal of Gin-Seng, & are persuaded that this plant has the virtue to make women fertile," he wrote.[123] In Louisiana, Jean Prat, correspondent to the Académie Royale des Sciences, had collected specimens of the "ginseng des natchès" from Jesuit missionaries and sent them on to the Jardin du Roi in Paris in 1736. From Québec, the root was sent to French apothecaries. The nun Marie-Andrée Regnard Duplessis, for example, exchanged American ginseng (along with beaver kidney, fir gum, and moose foot) with the Dieppe-based apothecary Jacques-Tranquillain Féret throughout the 1730s and 1740s.[124] Jean-François Gaultier recorded sending the plant to a physician in the Marine at Rochefort, where the effects of the root were being observed.[125]

In his effort to thoroughly describe indigenous and Chinese botanical practice, Lafitau had provided a virtual how-to guide for would-be aboriginal and Euro-American ginseng traders in his *Mémoire*. Lafitau equipped his readers with the language necessary to enlist a broad spectrum of collectors from the indigenous communities of colonial North America, even if he cautioned those who had not spent as much time among indigenous people as he had.[126] He provided clues not only to where ginseng might be found but to who might be able to find it and how to communicate with them. As collectors, merchants, and travelers expanded the known range of American ginseng in the decades after the publication of Lafitau's *Mémoire*, indigenous knowledge and informants were relegated increasingly minor roles in their texts. When the planter William Byrd II investigated the possibility of finding the plant in British North America while composing his *History of the Dividing Line betwixt Virginia and North Carolina*, he compared the Chinese, American, and South African locations where this plant was thought to be found.[127] Byrd's work took place in the context of broader Anglo-Atlantic study of ginseng that followed the spread of word of Jartoux's and Lafitau's work. Sébastien Vaillant, for example, sent copies of Lafitau's *Mémoire* to his English colleagues, including the botanist William Sherard.[128] Byrd saw little of value in either the Haudenosaunee or Khoikhoi knowledge that he glossed, equating and effacing them as providing only local names and evidence of the universal "esteem" for the plant.[129] Texts such as these could even remove the need for Native informants. In Louisiana, Jean-Baptiste Le Moyne de Bienville reported the ability to discover the plant even where local peoples had sought to keep their medical knowledge secret. In 1726, he wrote about the peoples along the Yazoo

River that "it is only ginseng and dittany that they have not been able to hide from us."[130]

The tools that Lafitau and Jesuits in China created helped make the comparative study of non-European cultures possible, but even as later authors engaged with Lafitau's *Mémoire*, they remained far more focused on understanding its Chinese contexts. Many texts that acknowledged Lafitau's discovery summarized it briefly, appending it to larger tracts that dealt with the plant's use and value in Asia. A 1722 article in the *Journal des sçavans*, for example, agreed that Lafitau had found ginseng in Canada but neglected to mention any Iroquoian knowledge about the plant "highly esteemed by the Chinese."[131] The 1725 *Dictionnaire universel* described ginseng as a plant that was "worth eight times its weight in silver" in China where it was "a sovereign remedy" before adding simply that "the Father Lafitau found this precious plant in Canada."[132] A dissertation on the use of ginseng by Jean-François Vandermonde presented and discussed in 1736 provided one of the most extensive examinations of the medicinal properties of the plant but also relied upon the knowledge of the "peoples of Asia, who from time immemorial have employed Ginseng, as a remedy proper to almost all maladies." About Iroquoian peoples, however, the author was silent, recounting only that Lafitau had found the plant in Canada.[133]

Each of these articles focused more on the prices cited by Lafitau and Jartoux than they did on the context of the Canadian discovery. Statements that the plant was used in lieu of taxes, was sold for silver, or was worth eight times its weight in silver were frequent. Reports of the price of ginseng in medical treatises, in dictionaries, and in scientific and commercial tracts were frequently stripped of their ethnographic context and forwarded as evidence of a commercial opportunity. A crucial threshold in interest seems to have passed in the 1730s when the Jesuit Jean-Baptiste du Halde was the first to suggest that the plant was worth its weight in gold, instead of silver. While recycling statements about the value of the plant in silver in his 1735 *Description géographique, historique, chronologique, politique, et physique de l'empire de la Chine et de la Tartarie chinoise*, du Halde added "this was true in the past; but now it is sold at the weight of gold" as a note at the bottom of the page.[134] Other authors soon quoted this price. Already by 1738 the English merchant and botanist Peter Collinson had written to Hans Sloane, then president of the Royal Society, about "The Ginseng a root so Celebrated for its vertues in China that it is Exchanged for Its weight in Gold," which had been found in the American colonies.[135] This price was noted in a variety of other texts in the decade after.[136]

These authors were ambivalent about whether the two ginsengs were the same plant. Instead, texts such as these read Jesuit and scientific accounts critically and with an explicit eye toward the development of an export economy to China. The author of the article "Gin-seng" in Diderot's *Encyclopédie*, for example, would cautiously state only that "all seem to presume that the two ginsengs are the same plant."[137] This author wrote, "In the end, without the need to seduce the Chinese with any preparation, it is certain that they do not know how to distinguish natural and pure ginseng from Canada from that of Tartarie: our Compagnie des Indes profits from their error, skillfully selling them one for the other, and has held the secret to this day to send three or four thousand pounds of ginseng from New France to China."[138] These comments were backed by scientific examination of the dyeing of roots in 1740 that claimed to demonstrate that many of the particular cues valued by Chinese consumers were manufactured.[139] For many, the only relevant litmus test was the acceptance of the Chinese market. If Lafitau had looked to Asian and American knowledge as incomplete but perfectible, the data that he and other Jesuits had assembled were now employed to manipulate. As the author of the entry on ginseng in the 1742 edition of the *Dictionnaire universel de commerce* explained: "the Chinese, at least most of them, are very ingenious in inventing new ways to make money. Above all, the illusion that some among them have about hidden virtues of natural things, & of the body of man as it relates to the usage that one can make in health and in sickness, often gives place to the spirit of charlatanism, which reigns strongly among them, to profit from it, to subtly find their advantage."[140] Esteem therefore became a symptom of an "avidity" that could be taken advantage of by educated merchants.[141] American and Eurasian flora could be manipulated to become identical, but the question about whether more essential transformations could be made— whether New World flora was an uncultivated variety of those known to Old World cultures—was left unstated and unresolved.

The growing ginseng trade encouraged indigenous peoples to travel further afield to harvest the plant. The traveling botanist Pehr Kalm wrote that the aboriginal peoples of the Saint Lawrence Valley were increasingly enlisted to provide French merchants with the valuable roots.[142] Already by the late 1740s, Kalm commented that "the Indians traveled about the country in order to collect as much as they could together, and to sell it to the merchants at Montréal."[143] The seasonality of the trade meant a conflict between agriculture and harvesting, so that "the French farmers were not able during that time to hire a single Indian, as they commonly do to help them with the harvest."[144] The trade in ginseng had a similar effect on indigenous communities in British North America. J. Martin Mack, a Moravian minister who traveled in Iroquoian territories in 1752, noted villages empty of all but those who could not participate in the harvest.[145] The trade in fact seemed to undermine imperial borders more generally, as a contraband trade and escalation in prices encouraged traveling between New France and New York. During engineer and surveyor Louis Franquet's stay at the Abenaki mission at Bécancour in 1753, he noted that "all the *Sauvages* had left to trade in New England, or were collecting geinseing; all the cabins were closed, in a league all that remained were those for whom infirmities or age impeded their ability to walk."[146] Missionaries even joined Natives as they traveled to collect and sell the plant in New York.[147] The trade brought in blankets and other important goods but also brought in liquor that, at least in the eyes of missionaries and administrators such as William Johnson, was detrimental to indigenous communities.[148]

The trade had a footprint all its own. By the mid-eighteenth century, for example, administrators such as the governor the marquis Ange Duquesne de Menneville and Intendant François Bigot were highlighting the fire hazard that ginseng collectors posed as they searched the woods.[149] The engineer Louis Franquet recorded burned bridges during his 1752 trip to Canada and suggested that "these accidents arrive most commonly through the inattention of the *sauvages*, the hunters and the geinseng seekers who fail to put out the fires that they have custom to make for their needs."[150] As colonists joined the trade they brought additional pressure on the populations being harvested. A 1752 ordinance in New France hinted that at the height of the trade, the lands of colonists and aboriginals alike were being invaded by would-be collectors.[151]

Lafitau wrote that the plant was already rare in New France, and William Byrd II wrote that the plant appeared "as sparingly as Truth & Publick Spirit" in Virginia.[152] Yet the numbers of plants collected from the plant's narrow range within New France grew enormously with the ginseng boom. Already by 1744, we can estimate that 171,600 to 300,300 plants were being collected to supply the legal trade to La Rochelle (see Table 1). By 1751, the number of plants collected had increased tenfold (1,859,000–3,253,250), before increasing almost another fourfold within a year at the height of the trade in 1752 (7,607,600–13,313,300). While this is lower than later recorded harvests (approximately 64 million roots were collected across the eastern United States

Table 1. Size and Value of the Legal Ginseng Trade at La Rochelle: American
Ginseng Imported from New France Annually, 1744–57

Year	Weight	Value	Plants (Low)	Plants (High)
1744	780	2,340	171,600	300,300
1747	2,150	12,900	473,000	827,750
1748	2,520	10,125	554,400	970,200
1749	5,450	76,300	1,199,000	2,098,250
1750	4,683	65,562	1,030,260	1,802,955
1751	8,450	152,100	1,859,000	3,253,250
1752	34,580	484,120	7,607,600	13,313,300
1753	35	140	7,700	13,475
1754	8,140	32,560	1,790,800	3,133,900
1756	568	1,420	124,960	218,680
1757	5,000	10,000	1,100,000	1,925,000

Source: Adapted from Rénald Lessard, "Pratique et praticiens en contexte colonial: le corps
médical canadien aux 17e et 18e siècles" (PhD diss., Université Laval, 1994), 192–93.
Note: Weight and value are in livres. Weights were converted from historical to modern
pounds at a ratio of 1.1:1. See Lester A. Ross, *Archaeological Metrology: English, French,
American, and Canadian Systems of Weights and Measures for North American Archaeology*
(Ottawa: National Historic Parks and Sites Branch, 1983). State and university agriculture
departments invested in the collection and conservation of American ginseng provide estimates
of between 200 and 350 plants per dried pound. See "Ginseng," *Cooperative Extension Service,
University of Kentucky, College of Agriculture* (2002), http://www.uky.edu/Ag/CDBREC
/introsheets/ginsengintro.pdf. For an estimate of 300 plants per pound, see "Ginseng Seasons
opens Sept. 1," *West Virginia Department of Commerce*, http://www.wvcommerce.org/news
/story/Ginseng_Season_Opens_Sept_1/2736/default.aspx. I have provided both low (200) and
high (350) estimates of the number of plants. Dried ginseng requires three pounds of fresh
roots, for a loss of around 66–70 percent of total weight in the drying process. See Andy
Hankins, "Producing and Marketing Wild Simulated Ginseng in Forest and Agroforestry
Systems," *Virginia Cooperative Extension, Virginia Tech/Virginia State University* (2009),
http://pubs.ext.vt.edu/354/354-312/354-312.html; Robert L. Beyfuss, "Ginseng Growing," *New
York State Department of Environmental Conservation*, http://www.dec.ny.gov/animals/7472
.html; idem, "Economics and Marketing of Ginseng," *USDA Forest Service/USDA Natural
Resource Conservation Service* (1999), http://www.wvfa.org/pdf/sfi/ginseng.pdf.

in 1841 alone, for example), the small possible range of collection in New France
suggests intense pressure on a geographically limited area.[153] Lafitau warned
against a repeat of the extinction of the plant that he assumed had occurred
in Europe millennia earlier; both ecological factors that limited the reproduc-
tion of the plant that he had observed in New France and an insatiable demand for its remarkable properties made this outcome likely.[154] He argued that
"the plant will soon be destroyed near the French habitations, and it will be
necessary to travel still further into the woods to search for it, which will make

it rare and very valuable."[155] Although severely overharvested populations have been shown to be able to recover, in the long term it seems likely that the ginseng trade limited the access of the Mohawk women of Kahnawake and other aboriginal communities to what Lafitau wrote had been "one of their ordinary remedies" before his discovery.[156]

In 1750 Pierre Poivre, botanist, missionary, and future intendant of Isle de France (present-day Mauritius), optimistically wrote that "it appears that the Chinese merchants have developed a taste for this merchandise in such a way that they are not soon to lose."[157] In the wake of Lafitau's *Mémoire*, the Jesuits' hopes for a new source of revenue for the colony seemed vindicated as the Compagnie des Indes Occidentales began shipping American ginseng to Chinese ports via La Rochelle. The profits were impressive: ginseng that could be bought for 3 francs per pound in New France commanded 180 francs in China, representing a 3,000 percent markup that was more than enough to cover the costs of the long transport.[158]

In 1751, the trade in ginseng was worth over 750,000 francs to the Compagnie des Indes.[159] Yet after 1752 the trade suffered as merchants and collectors neglected the plant's cultural contexts that had been a primary focus of Lafitau's text. More roots were collected in later years, but merchants had difficulty selling them.[160] The value of American ginseng was undermined as plants were picked out of season and were improperly processed, driving Chinese consumers away from the American plant. The Abbé Raynal described the ginseng trade as a paradigmatic example of colonial greed and the inefficient exploitation of what might have been a sustainable resource. In his 1770 *Épices et produits coloniaux*, Raynal wrote that "it finished itself in 1752 for five hundred thousand francs. The eagerness that this plant excited pushed the Canadians to collect in the month of May that which ought to have been collected in September, & to dry in the oven that which must be dried slowly in the shade."[161] Ignoring the information provided by Jartoux, Lafitau, and other Jesuit authors about the visual characteristics demanded by the Chinese market, colonial merchants bought plants that had been collected early and improperly cured, giving them an off color. Based upon the work of Jesuits who had written in China, Lafitau had warned that "when one pulls it out of the ground, it is necessary to wash it carefully, cut the root along its length so that it dries more easily. It is better to dry it in the shade than in the sun or by the fire, and store it in a dry place."[162] Yet colonial merchants compounded their error in collecting the plant early by neglecting to follow the process that gave ginseng root its trademark near transparency that Chinese consumers demanded. In China, where earlier unscrupulous Chinese merchants had mixed American and Chinese ginseng together to maintain prices and profit, American ginseng soon acquired a bad reputation as its difference became noticeable.[163] American ginseng was not much different from Chinese biologically, but through French inattention to the aesthetics of the plant in Asia, it was soon understood to be a distinct and inferior species by French and Chinese consumers.

*   *   *

If he had lived to see the ginseng boom, Lafitau might have seen New France more closely resemble the metropole. As Lafitau understood it, the same ginseng that was now being eradicated along the Saint Lawrence River had once also grown in France but had similarly been overharvested and driven to extinction. The Jesuit hoped that a transatlantic commerce in the plant could restore part of France's historical ecology, but his efforts had instead brought American environments into contact with commercial networks far more interested in extractive enterprise and a global commodities trade. He might have felt vindicated, however, that so many on both sides of the Atlantic remained committed to his botanical science that had focused on resemblance instead of difference and that had highlighted continuities and connections rather than drawing distinctions. While we might doubt that the merchants who shipped American ginseng to China in the 1740s and 1750s maintained the Jesuit's broader claims that one innocuous plant could reveal the contours of a deep and global history of movements and migrations, they clearly remained committed to proving biological (and commercially valuable) continuities between the Old World and the New.

The ginseng boom produced huge profits and ravaged colonial landscapes, yet it left larger questions about the nature of empire in New France unresolved. The story of American ginseng and its multiple discoveries reveals the extent to which, as the French regime in North America drew to a close in the eighteenth century, basic truths about the nature of New France's places and peoples were openly debated through the examination of American environments. Botanical claims were also political. Naturalists and botanists—figures we can confidently and anachronistically name "scientists"—arrogated the right to know the nature of New France better than those who lived there. Their science of difference challenged figures such as Lafitau whose vision of empire remained based on the assumption that the Americas and Europe were fundamentally similar. Lafitau's colonialism was patriarchal and intervention-ist, but it differed substantially from the extractive vision of French mer-chants or the understanding of New France as a biologically distinctive place developed by scientists.

This was not yet a "dispute of the New World," such as that in which figures such as Thomas Jefferson, Cornelius de Pauw, and Georges-Louis Leclerc, Comte de Buffon, engaged in the latter half of the eighteenth century. It was not yet an explicit question of whether the Americas were an inferior conti-nent, as Buffon and others would soon claim. Yet it was nonetheless a debate about how to frame the history of northeastern North America in a global context. Lafitau's claim that ginseng was a singular and globally distributed plant was also an argument for a single, providential human history in which both European and Iroquoian peoples shared a common origin. Read properly, ginseng was evidence that this New World was in fact neither newer nor es-sentially different than Europe. Less explicitly but no less powerfully, the sci-entists of the Académie who identified biologically distinct botanical species made a compelling case for an essential American difference. Even as the sun was beginning to set on France's American empire and political and military challenges to French rule grew, this epistemological crisis yawned wider still.

Cultivating New Relationships

"Canada produces almost no fruit," or so claimed Louis-Antoine, Comte de Bougainville, in 1757.[1] Anybody who has stepped outdoors in northeastern North America recognizes this statement for the absurdity that it is, and, as he continued in a *mémoire* "sur l'état de la Nouvelle France," Bougainville modified his statement. "Canada produces almost no fruit," he explained, "save for many types of admirable apples, principally *renettes, calvilles* and *api.*" He continued, "The most beautiful fruit is at Montréal, in the orchards of *Messieurs de Saint-Sulpice*, pears, lots of strawberries, raspberries and cherries, melons, very bad nuts that grow near Niagara, mediocre chestnuts, and a little *sauvage* fruit called *otaka* with which one makes preserves that would be delicious in France."[2]

As we follow the progression of his description, we can see that Bougainville was not saying that there were no fructifying plants native to North America. In general, in fact, his description of the colony's environment seemed fairly positive. Yet when he explained that "almost all the vegetables and herbs in France grow well there," Bougainville provided a clear sense of the perspective from which he assessed the American environments he experienced during the Seven Years' War.[3] It was as much aesthetic as it was botanical; fruit meant recognizably European spaces such as the orchards of Montréal and European plants grafted and transplanted into American soils. If it was a dismissal, then—and I think we ought to understand it as such—it was a dismissal of both indigenous plants and indigenous places and a rejection of any sort of relationship between them and the colony that he was actively trying to defend from British armies and influence.

Bougainville's sweeping claims are easy to contest, but we should not overestimate the coherency of his condemnation of American flora. The Comte's worldview was as complicated and, at times, as seemingly contradictory as many of the Enlightenment's engagements with its "others."[4] Canada possessed no fruit, for example, but there were "many rare plants of which the *Sauvages* know the properties very well."[5] We might compare Bougainville to Pehr Kalm, who, in a similar vein, was both confident enough to declare that as late as 1749 "not a single botanist had yet researched or carefully described the plants that are found" in North America and open about his reliance on colonial and indigenous communities for their knowledge.[6] Bougainville's writings did not actually clear the land of the many species of native fruits that continued—and continue still—to grow in northeastern North America, but they did signal a profound shift in the place that these *sauvage* plants occupied in imperial imaginings of New France.

In a few short lines, Bougainville signaled a definitive end to the optimism that had fueled seventeenth-century efforts to reshape both the natural and moral landscapes of French North America. Privileging a discussion of plants imported from Europe, he neglected to mention many plants that, only a century earlier, were targeted in French plans to transform North American plants to more perfectly resemble their European analogues. His summary reflected a growing sense of a more essential difference between the floras of New France and Old that persisted in spite of considerable success in introducing the botanical cornerstones of French life to gardens and fields across the continent. It seems little wonder that, within a few short years, France would opt to keep Guadeloupe—a carefully cultivated, thoroughly manipulated, and socially and biologically engineered landscape—instead of Canada during the negotiations at the end of the Seven Years' War.[7]

Bougainville's sense of what counted as cultivated plants and places in North America was, then, far more limited than that of the colonists, missionaries, and explorers who had founded New France a century and half earlier. This book has followed these other understandings of cultivation to trace the evolving relationship between knowledge about nature *in* New France and the nature *of* New France between the founding of the colony in the seventeenth century and its fall to the British in the eighteenth. Cultivation provided a powerful metaphor and a set of practices through which French colonists, administrators, missionaries, and explorers engaged with and produced knowledge about American environments and people. When authors such as Samuel de Champlain first spoke about cultivating a New France in North America, they transplanted an ideology then intimately associated with the reassertion of royal and patriarchal authority across France. Championed by a new generation of agricultural reformers and political theorists, it valorized the engagement with a diverse cast of agencies, as often human as they were non-human. As such, it suggested the need for an outward-looking and empiricist method of knowledge production; its promised land was less a landscape swept clean of recalcitrant matter than it was a well-choreographed harmony of otherwise discordant forces brought together under the supervision of (and in celebration of) a benevolent patriarch. The uncultivated plants made invisible within Bougainville's sweeping denunciation were then specifically those that had interested earlier generations of colonists, administrators, explorers, and missionaries.

At the dawn of French colonialism in the Saint Lawrence Valley, figures such as Champlain described environments that were *sauvage* and that demanded the intervention of French colonists who took up the civilizing mission that had been left by the Romans who had civilized their Gallic ancestors. It was an ideology that effaced indigenous ecological knowledge and that collapsed distinctions between indigenous-created environments and the uninhabited places of North America. Yet their focus on cultivation encouraged colonists to actively engage American environments, and they came, through their experience, to appreciate the distinctiveness of these new places and cultivated ecological and intellectual spaces in which they interacted with indigenous plants and peoples. As the scale of French ambitions decreased, calls for the cultivation of *sauvage* plants gave way to efforts to more fully transplant French ecological regimes in Canada and supported administrative efforts to create boundaries between American and French spaces. Newly emergent botanical sciences seemed to give colonial administrators and authors new tools with which to make sense of the increasingly foreign environments in which they lived. The complex body of knowledge about North American environments that had been acquired through a century of encounter, however, proved difficult to reconcile with scientific methods that allowed little room for considerations of place or for the knowledge of colonial and indigenous peoples.

Bougainville wrote in a decade during which European authors felt confident in characterizing the whole of the Americas in sweeping generalizations about the continents' geography and environments. Observable biological differences were broadly pathologized during this period. Race, to name the most frequently studied example of this trend, was a newly visible concept in this period, redefined and redeployed in the work of Enlightenment authors who sought to produce "a unified science of man" that was nonetheless varied in its methods and arguments.[8] In 1749, writes literary scholar Andrew Curran, the work of Georges-Louis Leclerc, the Comte de Buffon, "ushered in what would become a new era in the interpretation of the African," but this was but one point in a far larger reconceptualization of biological diversity during the eighteenth century.[9] Buffon was therefore a leading participant in a broader movement that shifted discussions of race away from religious and noble contexts and toward an identification with the methods and epistemologies of a renovated natural history.[10] Within this schema, individual races were deviations from an ideal type, inevitably understood to be that which existed in the Old World and, more specifically, in Europe itself.[11] Outside of Europe, then, natural historians were presented with examples of degenerations rather than truly novel types.

Buffon's language seems reminiscent of authors such as Champlain and Boucher, those who suggested that mutable differences might be cultivated out of American flora and peoples. Yet beneath this similarity in language and a shared sense of the essential mutability of biological types, differences mount. Buffon sought to explain the novelty that was increasingly observed by authors such as Bougainville by pointing to more essential differences between the continents—determined by their environments and, in the end, by their relative geological age.[12] The New World, for example, was just that. It had more recently emerged from the oceans and was therefore more humid and more tropical.[13] In the writings of Buffon, the living things who resided there were therefore subject to agencies that could not be cultivated out through the efforts of colonists, nor could the environment of the continent itself be remade through colonialism.[14] Where cultivation had encouraged colonial authors to study and record their encounters with non-human agencies, Buffon's natural history catalogued the elements of irreducibly foreign environments.[15]

New France occupied (and occupies still) an ambiguous place within this intellectual history. Many Americans know that Thomas Jefferson thought that a moose specimen would offer a specific rebuke to the generalizations of European intellectuals.[16] Canadians know all too well that other Enlightenment authors were ready to offer their own sweeping generalizations about France's northern American colonies. Voltaire of course most famously stated that "I love peace much more than I love Canada" and that the entire colony was little more than "a few acres of snow."[17] The authors at the heart of the Enlightenment's study of human and biological diversity—men such as Diderot, Buffon, and Rousseau—read travel literature and other accounts written from New France.[18] Yet those who have looked closer have demonstrated that even the most widely read and influential texts from New France were misread or ignored even as they maintained a broad cultural importance.[19] Nonetheless, this study of cultivation suggests that these better-known intellectual debates were anticipated and informed by more than a century of conflict over the significance of colonial experience and over the methods through which knowledge of the New World could be produced.

By the end of the French regime in New France, engagement had turned to retrenchment; cultivation as conversion had transitioned to cultivation as replacement. New France was more new than it would ever be French. Yet if we are tempted to valorize older forms of knowledge production that seem more attentive to alternatives, more just to indigenous peoples, and more willing to accept ecological and cultural difference, we should understand that, for the various French communities studied in this book, knowing colonial environments meant projecting colonial relations onto the continent's indigenous peoples and places. Discourses about indigenous and colonial ecologies most often served to legitimate the appropriation of land and the cultural and geographical displacement of aboriginal communities. The ideologies of the founders of French colonialism in North America then anticipated—even if they do not neatly map onto—the forms of settler colonialism identified by, among others, scholars such as Patrick Wolfe and Lorenzo Veracini.[20]

We might consider, for example, the discourses of the colonial promoters throughout the seventeenth century as effecting a "transfer" or "collective sovereign displacement" in colonial North America.[21] Lescarbot's historical narratives situated his colonial present and his compatriots' labors within a teleological timeline of French colonial expansion.[22] The providential narratives of Jesuit and Franciscan missionaries anticipated the political and spiritual integration of Native peoples into French colonial life but explicitly called for the displacement of indigenous cultures and the relocation of Native communities.[23] Pierre Boucher's description of Laurentian landscapes effectively named Haudenosaunee occupations of fertile lands to the south unnatural and called for their aggressive displacement.[24] In each case, these and the other authors considered in this book imagined the inevitable replacement of indigenous cultures and ecologies and obscured the unalterable fact that this was an invasion.[25] The violence of their encounters with the land and aboriginal peoples was elided by discourses that promised that these were only short-term pains on the path to the lasting peace of domestication.[26] Although the violence of pruning, grafting, clearing, and cultivating might seem trivial alongside documented attempts at genocide such as the eighteenth-century

Fox Wars, we should nonetheless see these as related manifestations of a broader impulse aimed at subduing indigenous agencies for their effective replacement.[27]

Recognizing the violent impulses that lay at the heart of the seemingly innocuous colonial study of plants demonstrates the entanglement of ecological and political histories that might otherwise be told separately. The language of cultivation embraced by French colonialism allows us to appreciate the scale of its ambitions; a colonial project most often studied through its effects on peoples equally targeted the region's non-human inhabitants. The history of colonial science, a field that has clearly demonstrated the extra-European origins of much of what became Enlightenment natural history in the past decade, seems, however, insufficient to the task of writing the history of actors who understood their own place in natural and social worlds. Instead, the scale of French colonial ambition suggests the need for the methodologies and theoretical commitments of scholars of political ecology who have fore-grounded the entanglements of "nature, culture and power, and politics, broadly speaking."[28] Put more simply, to tell these stories we need to "ecologise [our] politics."[29]

Taking up this challenge is made easier by the work of indigenous and Native American studies scholars who have demonstrated the frustrating longevity and violent legacy of colonial political ecologies that discursively and physically displaced the traditional ecological knowledge of indigenous peoples in North America. Tragic histories of allotment in the United States and "peasant" agricultural programs in Canada demonstrate the continuation of efforts to use agriculture to assimilate indigenous peoples.[30] In the Pacific Northwest, the U.S. government placed restrictions on indigenous uses of their oceans, even as the commercial exploitation of whales and other marine species dramatically undermined their survival.[31] In the twentieth century, environmental sciences continued the construction of indigenous ecological knowledge as irresponsible and outmoded; adaptation to new ecological and economic realities delegitimated indigenous knowledge and conservation sciences saw indigenous communities as threats to ecological processes.[32]

More recently still, American manifestations of the global phenomenon of biopiracy have continued to invoke colonial discourses that map differences between cultivated and wild plants onto unspoken civilizational deficiencies of indigenous peoples. Winona LaDuke, for example, has written extensively about the inability of Ojibwe people to contest the patenting of wild rice, a crop that remains integral to Great Lakes cultures whose contributions to the evolution of the plant are obfuscated behind a discourse of the "natural" state of plants that have not been selectively bred or genetically engineered by Euro-American corporations.[33] These stories offer stark lessons of the power of discourses such as cultivation to continue, even as the places that they called into being—colonial gardens, orchards, and New France itself—disappeared.

To confront Bougainville's stark vision of a Canada without fruit is therefore to take up the challenge of cultivating new relationships with North America's indigenous peoples and places. It is an ancillary to the project of Native and non-Native scholars who have actively sought to decolonize traditional ecological knowledge and mobilize it to confront the pressing cultural and ecological threats that centuries of aggressive agricultural expansion have wrought.[34] It demands not a return to the ecological vision of Champlain but an awareness of our own connectedness, of our own entanglement in environments that are political and polities that are also ecological.

NOTES

INTRODUCTION

1. Samuel de Champlain, *The Works of Samuel de Champlain*, ed. Henry Percival Biggar (Toronto: University of Toronto Press, 1971), 2:52.

2. Brian Brazeau, *Writing a New France, 1604–1632: Empire and Early Modern French Identity* (Farnham: Ashgate, 2009); Catherine Ferland, *Bacchus en Canada: Boissons, buveurs et ivresses en Nouvelle-France* (Sillery: Septentrion, 2010); Sara E. Melzer, *Colonizer or Colonized: The Hidden Stories of Early Modern French Culture* (Philadelphia: University of Pennsylvania Press, 2012).

3. Olivier de Serres, *Le théâtre d'agriculture et mesnage des champs*, 2nd ed. (Paris, 1603), 127.

4. For a narrative overview of these months, see David Hackett Fischer, *Champlain's Dream: The Visionary Adventurer Who Made a New World in Canada* (Toronto: Alfred A. Knopf Canada, 2008), 227–53.

5. A full examination of these early years and decades of exploration and settlement is beyond the scope of this introduction and book. Readers interested in these broader contexts should consult Éric Thierry, *La France de Henri IV en Amérique du Nord: De la création de l'Acadie à la fondation de Québec* (Paris: Honoré Champion, 2008); Marcel Trudel, *Histoire de la Nouvelle France, Le Comptoir, 1604–1627* (Montréal: Fides, 1966).

6. Champlain, *Works*, 2:44.

7. Ibid., 2:52.

8. Ibid., 2:147.

9. Marc Lescarbot, *Histoire de la Nouvelle-France: Contenant les navigations, découvertes, & habitations faites par les François aux Indes Occidentales & Nouvelle-France* (Paris: Adrian Perier, 1617), 474.

10. Gabriel Sagard, *Le Grand Voyage du pays des Hurons: Suivi du Dictionnaire de la langue huronne*, ed. Jack Warwick (Montréal: Presses de l'Université de Montréal, 1998), 146. Place-names such as "New France" and "Canada" pointed to shifting targets during this period. I will follow historical actors in their definition, however, which means that while the beginning of this book will include what came to be known as Acadia, the focus will shift west to include debates about the relationship between this New France and newly constituted spaces such as the *pays d'en haut*, Illinois, and Louisiana. I will identify "Canada" when it is the term used by historical actors and will otherwise favor the less overtly presentist "America" or, as an adjective, "American." See Catherine Desbarats and Allan Greer, "Où est la Nouvelle-France?" *Revue d'histoire de l'Amérique française* 64, no. 3–4 (2011): 31–62.

11. Sagard, *Le Grand Voyage*, 147.

12. Ibid.

13. Gabriel Sagard, *Histoire du Canada et voyages que les frères mineurs Récollects y ont faicts pour la conversion des Infidelles* (Paris: Chez Claude Sonnius, 1636), 162. See also Daniel Simoneau, "The Seminary of Québec Site: From New France's Earliest Farm to Its First Religious Institution," *Post-Medieval Archaeology* 43, no. 1 (2009): 213–28.

14. The best overview of the expansion of gardens during this period is Marie-Josée Fortier, *Les jardins d'agrément en Nouvelle-France: Étude historique et cartographique* (Québec: Les Éditions GID, 2012). For studies of agricultural expansion, see Jacques Guimont, *La Petite-ferme du cap Tourmente, un établissement agricole tricentenaire: De la ferme de Champlain aux grandes volées d'oies* (Québec: Septentrion, 1996), ch. 1; Serge Courville, *Le Québec: Genèses et mutations du territoire* (Sainte-Foy: Presses de l'Université Laval, 2000), ch. 4.

15. An archaeological study of Champlain's *habitation* found seeds of many local plants. Catherine Fortin, "Appendice no. 3: Étude de grains provenant du site de l'Habitation de Champlain, maison Marquis," in *L'Habitation de Champlain*, ed. Françoise Niellon and Marcel Moussette (Québec: Gouvernement du Québec, Ministère de la Culture et des Communications, 1985), 374.

16. Champlain, *Works*, 2:147.

17. Champlain's study of American landscapes is discussed in greater detail in Chapter 1. See also Christian Morissonneau, "Le Nouveau Monde: Les perceptions et representations de Champlain," in *Le Nouveau Monde et Champlain*, ed. Guy Martinière and Didier Poton (Paris: Les Indes savantes, 2008), 43–52.

18. Champlain, *Works*, 2:60–61.

19. Ibid., 2:257.

20. Ibid., 2:23.

21. Jill H. Casid, *Sowing Empire: Landscape and Colonization* (Minneapolis: University of Minnesota Press, 2005), ch. 1.

22. This has most influentially been done by Olive Patricia Dickason, *The Myth of the Savage, and the Beginnings of French Colonialism in the Americas* (Edmonton: University of Alberta Press, 1984). See also François Marc Gagnon, *Ces hommes dits sauvages: L'histoire fascinante d'un préjugé qui remonte aux premiers découvreurs du Canada* (Montréal: Libre Expression, 1984). For a broader Atlantic comparison, see Frank Lestringant, *Le Huguenot et le sauvage: L'Amérique et la controverse coloniale en France, au temps des guerres de religion (1555–1589)* (Paris: Klincksieck, 1990).

23. Sophie White, *Wild Frenchmen and Frenchified Indians: Material Culture and Race in Colonial Louisiana* (Philadelphia: University of Pennsylvania Press, 2012), 235. Gordon Sayre notes that the different European-language names used to describe indigenous peoples are best viewed as "representing different theories that Euramericans had and have about the origin and status of the natives of America." Gordon M. Sayre, *Les Sauvages Américains: Representations of Native Americans in French and English Colonial Literature* (Chapel Hill: University of North Carolina Press, 1997), xv.

24. Dickason, *The Myth of the Savage*, 63.

25. ARTFL, *Dictionnaires d'autrefois*, http://artfl-project.uchicago.edu/content/dictionnaires-dautrefois.

26. The most sustained analysis of the differences between the French *sauvage* and English "savage" can be found in Thomas G. M. Peace, "Deconstructing the Sauvage/Savage in the Writing of Samuel de Champlain and Captain John Smith," *French Colonial History* 7, no. 1

(2006): 1–20. See also Melzer, *Colonizer or Colonized*, 87; John Duval and Kathleen Duval, "Are Sauvages Savages, Wild People, or Indians in a Colonial American Reader?" *Translation Review* 79, no. 1 (2010): 1–16; Nancy Senior, "Of Whales and Savages," *Meta* 49, no. 3 (2004): 462–74.

27. Champlain, *Works*, 2:60–61.

28. This partook in a broader shift in the theorization of education in the French Atlantic world. In France, education helped "replace military attributes with elements of social polish and 'culture' that increasingly made nobility a social status accessible to an elite recruited from a variety of professional groups." Mark Edward Motley, *Becoming a French Aristocrat: The Education of the Court Nobility, 1580–1715* (Princeton, N.J.: Princeton University Press, 1990), 4.

29. Serres, *Le théâtre d'agriculture et mesnage des champs*, preface.

30. Sara Melzer has explored the significance of this term more fully in *Colonizer or Colonized*, 204–9.

31. Recent scholarship has highlighted the centrality of assimilation to France's colonial project in North America but has also emphasized the extent to which this remained an unrealized ambition. See, for example, Saliha Belmessous, "Assimilation and Racialism in Seventeenth- and Eighteenth-Century French Colonial Policy," *American Historical Review* 110, no. 2 (2005): 322–49.

32. "Improvement" is a term more often associated with later eras of colonialism in the Anglo-Atlantic world, but it was similarly tied to an emerging confidence in the ability to ameliorate the human and natural worlds with new knowledge. The seventeenth-century roots of this term have recently been explored in Vittoria Di Palma, *Wasteland: A History* (New Haven, Conn.: Yale University Press, 2014), ch. 2. See also Richard Drayton, *Nature's Government: Science, Imperial Britain, and the "Improvement" of the World* (New Haven, Conn.: Yale University Press, 2000); Paul Warde, "The Idea of Improvement, c. 1520–1700," in *Custom, Improvement and the Landscape in Early Modern Britain*, ed. Richard W. Hoyle (Farnham: Ashgate, 2011), 127–48.

33. Political ecology highlights the entanglement of natural and social worlds, the articulation and exercise of power within a settler colonial context, and the materiality of a subject that might otherwise be understood as an intellectual history. Political ecology "has focused primarily on the politics that surround environmental change, conservation interventions, and natural resource economies." Mara J. Goldman and Matthew D. Turner, introduction to *Knowing Nature: Conversations at the Intersection of Political Ecology and Science Studies*, ed. Mara J. Goldman, Paul Nadasdy, and Matthew D. Turner (Chicago: University of Chicago Press, 2011), 4. It has therefore been an approach favored more by scholars of the present than those of the past. It has recently become an important term for historians of the environment in the Atlantic world. See Molly Warsh, "A Political Ecology in the Early Spanish Caribbean," *William and Mary Quarterly* 71, no. 4 (2014): 517–48; Keith Pluymers, "Atlantic Iron: Wood Scarcity and the Political Ecology of Early English Expansion," *William and Mary Quarterly* 73, no. 3 (2016): 389–426. I have been particularly inspired by the science studies and new materialism scholars who have embraced the term to foreground the material histories of knowledge production and exchange. See Jane Bennett, *Vibrant Matter: A Political Ecology of Things* (Durham, N.C.: Duke University Press, 2009); Arturo Escobar, *Territories of Difference: Place, Movements, Life, Redes* (Durham, N.C.: Duke University Press, 2008); Mara J. Goldman, Paul Nadasdy, and Matthew D. Turner, eds., *Knowing Nature: Conversations at the Intersection of Political Ecology and Science Studies* (Chicago: University of Chicago Press, 2011); Anna Lowenhaupt Tsing, *Friction: An Ethnography of Global Connection* (Princeton, N.J.: Princeton University Press, 2005).

34. Rueben Gold Thwaites, ed., *The Jesuit Relations and Allied Documents*, 73 vols. (Cleveland: Burrows Brothers Company, 1896–1901), 4:111 (hereafter *JR*).

35. The centrality of these commodities to our understanding of this period owes a particular debt to the work of Harold Innis, as he focused on the role of these "staple" goods in colonial, imperial, and national economies. See Harold Adams Innis, *The Fur Trade in Canada: An Introduction to Canadian Economic History* (New Haven, Conn.: Yale University Press, 1930); idem, *The Cod Fisheries: The History of an International Economy* (New Haven, Conn.: Yale University Press, 1940).

36. My focus has been most inspired by those scholars who identify as historians of science rather than environmental historians. Victoria Dickenson has written, for example, that "many of the plants and animals were identical or very similar to those in northern Europe. The European response to the landscape, then, was not to see it as bizarre, but rather as not similar but at least recognizable." Victoria Dickenson, *Drawn from Life: Science and Art in the Portrayal of the New World* (Toronto: University of Toronto Press, 1998), 115. See also Alain Asselin, Jacques Cayouette, and Jacques Mathieu, *Curieuses histoires de plantes du Canada*, vol. 1 (Québec: Septentrion, 2014); idem, *Curieuses histoires de plantes du Canada*, vol. 2 (Québec: Septentrion, 2015). The scholarship on agriculture and the seigneurial system is extensive. For a recent introduction, see Benoît Grenier, *Bréve histoire du régime seigneurial* (Montréal: Boréal, 2012). Few historians of seigneurialism have identified their work as environmental history, even if their work necessarily pays close attention to landscapes of agricultural production. The notable exception is Colin Coates, whose scholarship aims to demonstrate the entanglement of nature and society in two Saint Lawrence seigneuries. See *Metamorphoses of Landscape and Community in Early Quebec* (Montréal: McGill-Queen's University Press, 2000).

37. Cole Harris, *The Reluctant Land: Society, Space, and Environment in Canada Before Confederation* (Vancouver: University of British Columbia Press, 2008). See also Ramsay Cook, *1492 and All That: Making a Garden out of a Wilderness* (North York: Robarts Centre for Canadian Studies, 1993). This is also how early American historians have characterized the region in broader comparative studies. See, for example, Alan Taylor, *American Colonies: The Settling of North America* (New York: Penguin Books, 2002), 369.

38. Frenchness (or *francité*) in a French colonial context has emerged as a productive area of scholarship in recent years but is primarily considered within the context of human societies. See Saliha Belmessous, "Etre français en Nouvelle-France: Identité française et identité coloniale aux dix-septième et dix-huitième siècles," *French Historical Studies* 27, no. 3 (2004): 507–40; Gilles Havard, "Les forcer à devenir Cytoyens," *Annales: Histoire, Sciences Sociales* 64, no. 5 (2009): 985–1018; White, *Wild Frenchmen and Frenchified Indians*. This has also been explored in a broader French Atlantic context but with a more explicit focus on the eighteenth century. See Doris Garraway, *The Libertine Colony: Creolization in the Early French Caribbean* (Durham, N.C.: Duke University Press, 2005); John D. Garrigus, *Before Haiti: Race and Citizenship in French Saint-Domingue* (New York: Palgrave Macmillan, 2006).

39. This is a conversation opened up in Bruno Latour, *We Have Never Been Modern* (Cambridge, Mass.: Harvard University Press, 1993), 1–12. For a discussion of the epistemological and ontological challenges of objects such as global warming—objects that Timothy Morton calls "hyperobjects" because of their intractability—see Timothy Morton, *Hyperobjects: Philosophy and Ecology After the End of the World* (Minneapolis: University of Minnesota Press, 2013).

40. On the etymology and significance of the word *sauvage* in French colonialism, see Dickason, *The Myth of the Savage*, 63–64.

41. See Anthony Pagden, *Lords of All the World: Ideologies of Empire in Spain, Britain and France c. 1500–c. 1800* (New Haven, Conn.: Yale University Press, 1995). This echoes studies that have shown the extent to which New England and Anglo-American colonialism more broadly were articulated in dialogue with histories of Iberian colonialism. See Ralph Bauer, *The Cultural Geography of Colonial American Literatures: Empire, Travel, Modernity* (Cambridge: Cambridge University Press, 2003); Jorge Cañizares-Esguerra, *Puritan Conquistadors: Iberianizing the Atlantic, 1550–1700* (Stanford, Calif.: Stanford University Press, 2006); Lisa Voigt, *Writing Captivity in the Early Modern Atlantic: Circulations of Knowledge and Authority in the Iberian and English Imperial Worlds* (Chapel Hill: University of North Carolina Press, 2009).

42. Lorenzo Veracini suggests that "transfer" is "a foundational trait of settler colonial formations" that legitimizes dispossession and the transfer of sovereignty. Lorenzo Veracini, *Settler Colonialism: A Theoretical Overview* (New York: Palgrave Macmillan, 2010), 17.

43. This has been foregrounded most effectively by scholars who have studied the footprint of African knowledge and slavery. See Jennifer L. Anderson, *Mahogany: The Costs of Luxury in Early America* (Cambridge, Mass.: Harvard University Press, 2012); Judith Ann Carney, *Black Rice: The African Origins of Rice Cultivation in the Americas* (Cambridge, Mass.: Harvard University Press, 2001).

44. This has been most amply studied in the Iberian Atlantic world. See Matthew James Crawford, *The Andean Wonder Drug: Cinchona Bark and Imperial Science in the Spanish Atlantic, 1630–1800* (Pittsburgh: University of Pittsburgh Press, 2016); Pablo F. Gómez, *The Experiential Caribbean: Creating Knowledge and Healing in the Early Modern Atlantic* (Chapel Hill: University of North Carolina Press, 2017); Timothy D. Walker, "The Medicines Trade in the Portuguese Atlantic World: Acquisition and Dissemination of Healing Knowledge from Brazil (c. 1580–1800)," *Social History of Medicine* 26, no. 3 (2013): 403–31.

45. As historians of science in early modern Europe and the Atlantic world have shown us in elegant and influential studies, new approaches to the natural world in this period "emphasized practice, the active collection of experience, and observation of nature." Pamela H. Smith and Paula Findlen, "Introduction: Commerce and the Representation of Nature in Art and Science," in *Merchants & Marvels: Commerce, Science, and Art in Early Modern Europe*, ed. Pamela H. Smith and Paula Findlen (New York: Routledge, 2002), 3. See also Brian W. Ogilvie, *The Science of Describing: Natural History in Renaissance Europe* (Chicago: University of Chicago Press, 2006).

46. This is inspired by Jennifer Anderson's elegant evocation of the "sylvan alchemy" that transmuted mahogany from wild tree to valuable commodity. Anderson, *Mahogany*, 15.

47. For an overview on this oft-discussed topic in the history of science, see Lorraine Daston, "The Nature of Nature in Early Modern Europe," *Configurations* 6, no. 2 (1998): 149–72; Lorraine Daston and Katharine Park, *Wonders and the Order of Nature, 1150–1750* (Cambridge: Zone Books, 1998). The category of the "natural" was particularly confused in gardening where effects such as grottos and fountains intentionally played with its distinction from artifice. See Danièle Duport, *Le jardin et la nature: Ordre et variété dans la littérature de la Renaissance* (Geneva: Droz, 2002), ch. 2; and John Dixon Hunt, *Greater Perfections: The Practice of Garden Theory* (Philadelphia: University of Pennsylvania Press, 2000), ch. 3.

48. Antonio Barrera-Osorio, *Experiencing Nature: The Spanish American Empire and the Early Scientific Revolution* (Austin: University of Texas Press, 2006). See also Miguel de Asúa

and Roger French, *A New World of Animals: Early Modern Europeans on the Creatures of Iberian America* (Aldershot, UK: Ashgate, 2005); Drayton, *Nature's Government*; Antonello Gerbi, *Nature in the New World: From Christopher Columbus to Gonzalo Fernández de Oviedo*, trans. Jeremy Moyle (Pittsburgh: University of Pittsburgh Press, 2010). This is a position that historians of science in Europe have recently criticized. See Alix Cooper, *Inventing the Indigenous: Local Knowledge and Natural History in Early Modern Europe* (Cambridge: Cambridge University Press, 2007), 3; Ogilvie, *The Science of Describing*, 143.

49. For examinations of Buffon's theory about the novelty of the New World and the "dispute of the New World" that followed, see Lee Alan Dugatkin, *Mr. Jefferson and the Giant Moose: Natural History in Early America* (Chicago: University of Chicago Press, 2009); Antonello Gerbi, *The Dispute of the New World: The History of a Polemic, 1750–1900* (Pittsburgh: University of Pittsburgh Press, 1973).

50. In this I follow scholars who focus on the local histories of early modern and Enlightenment science that emerged through prolonged contact with indigenous cultures and American environments. Daniela Bleichmar, *Visible Empire: Botanical Expeditions & Visual Culture in the Hispanic Enlightenment* (Chicago: University of Chicago Press, 2012); Joyce E. Chaplin, *Subject Matter: Technology, the Body, and Science on the Anglo-American Frontier, 1500–1676* (Cambridge, Mass.: Harvard University Press, 2001); Kathleen S. Murphy, "Translating the Vernacular: Indigenous and African Knowledge in the Eighteenth-Century British Atlantic," *Atlantic Studies* 8, no. 1 (2011): 29–48; Susan Scott Parrish, *American Curiosity: Cultures of Natural History in the Colonial British Atlantic World* (Chapel Hill: University of North Carolina Press, 2006); Neil Safier, *Measuring the New World: Enlightenment Science and South America* (Chicago: University of Chicago Press, 2008); Londa Schiebinger, *Plants and Empire: Colonial Bioprospecting in the Atlantic World* (Cambridge, Mass.: Harvard University Press, 2004); Cameron B. Strang, "Indian Storytelling, Scientific Knowledge, and Power in the Florida Borderlands," *William and Mary Quarterly* 70, no. 4 (2013): 671–700.

51. If I do not consider the alternative literacies of indigenous peoples, which scholars have demonstrated contain a wealth of information about indigenous worldviews and knowledge systems, I do subject these familiar sources to techniques now used by generations of ethnohistorians to recover indigenous contributions to French knowledge. I have relied particularly heavily on the technique known as "upstreaming" that uses anthropological studies of contemporary or near-contemporary aboriginal cultures for insights into historical populations, first popularized by ethnohistorians such as William Fenton, Bruce Trigger, and Colin Calloway and an accepted practice for generations of scholars since. This approach might be criticized for an assumption of cultural uniformity through time, but it has provided a larger context in which to understand the traces of indigenous cultural practices that are found in colonial texts and a basis from which to understand how indigenous knowledge systems have interacted with those of Europeans and colonists over time. William Fenton writes that the "method rests on three premises. First, major patterns of culture tend to be stable over long periods of time. Second, one proceeds from what is known to examining sources that may contain familiar elements. Third, the ethnologist favors those sources that ring true ethnologically and resonate at both ends of the time span." William N. Fenton, *The Great Law and the Longhouse: A Political History of the Iroquois Confederacy* (Norman: University of Oklahoma Press, 1998), xvi.

52. Gerbi, *The Dispute of the New World*. See also Jorge Cañizares-Esguerra, *How to Write the History of the New World: Histories, Epistemologies, and Identities in the Eighteenth-Century Atlantic World* (Stanford, Calif.: Stanford University Press, 2001).

CHAPTER 1

1. Sagard, *Le Grand Voyage*, 136.

2. Stephen S. Birdsall et al., *Regional Landscapes of the United States and Canada*, 7th ed. (Hoboken, N.J.: John Wiley and Sons, 2009), 21.

3. Sagard, *Le Grand Voyage*, 136.

4. Marie-Victorin, *Flore laurentienne* (Montréal: Presses de l'Université de Montréal, 1964), 5–6.

5. "Biomes represent regions with characteristic combinations of plants defined based on their physiognomy rather than species identity, sometimes with climate or locality explicitly incorporated in their definition." Glenn R. Moncrieff, Thomas Hickler, and Steven I. Higgins, "Intercontinental Divergence in the Climate Envelope of Major Plant Biomes," *Global Ecology and Biogeography* 24, no. 3 (2015): 324. Moncrieff and colleagues also explain that, because of historical connections, this connection is deeper than many linkages that can be made between biomes that are geographically isolated (330).

6. Jacques Rousseau, "La forêt mixte du Québec dans la perspective historique," *Cahiers de géographie du Québec* 7, no. 13 (1962): 111–20.

7. Indigenous ecological practice intentionally increased the number of fruit and nut-bearing trees that colonial authors frequently sought out. Marc D. Abrams and Gregory J. Nowacki, "Native Americans as Active and Passive Promoters of Mast and Fruit Trees in the Eastern USA," *The Holocene* 18, no. 7 (2008): 1123–37. For a synthesis of the effect of indigenous cultures on North American environments, see Paul A. Delcourt and Hazel R. Delcourt, *Prehistoric Native Americans and Ecological Change: Human Ecosystems in Eastern North America Since the Pleistocene* (Cambridge: Cambridge University Press, 2004). For a broad argument for Native agency in American environmental history, see William M. Denevan, "The Pristine Myth: The Landscape of the Americas in 1492," *Annals of the Association of American Geographers* 82, no. 3 (1992): 369–385.

8. Sagard, *Le Grand Voyage*, 148.

9. Sagard has often been highlighted by scholars as particularly appreciative of Canadian nature. Lynn Berry, "The Delights of Nature in This New World: A Seventeenth-Century Canadian View of the Environment," in *Decentring the Renaissance: Canada and Europe in Multidisciplinary Perspective, 1500–1700*, ed. Germaine Warkentin and Carolyn Podruchny (Toronto: University of Toronto Press, 2001), 229; Réal Ouellet and Alain Beaulieu, "Avant-propos," in *Rhétorique et conquête missionaire: Le jésuite Paul Lejeune*, ed. Réal Ouellet (Sillery: Septentrion, 1993), 21.

10. Sagard, *Le Grand Voyage*, 312–14.

11. Ibid., 314.

12. Ibid., 313, 315.

13. Antonio Barrera-Osorio has even seen the origins of modern science itself in this conflict, while others have suggested that, at the very least, the experience of American novelty severely challenged the validity of classical sources that had long defined the flora and fauna of the Old World. See Barrera-Osorio, *Experiencing Nature*. See also, for example, Anthony Pagden, *European Encounters with the New World: From Renaissance to Romanticism* (New Haven, Conn.: Yale University Press, 1993); Anthony Grafton, *New Worlds, Ancient Texts: The Power of Tradition and the Shock of Discovery* (Cambridge, Mass.: Belknap Press of Harvard University Press, 1992); Drayton, *Nature's Government*, particularly ch. 1. This has been most explicitly

discussed in a botanical context in Michael T. Ryan, "Assimilating New Worlds in the Sixteenth and Seventeenth Centuries," *Comparative Studies in Society and History* 23, no. 4 (1981): 519–38.

14. Jonathan D. Sauer, "Changing Perception and Exploitation of New World Plants in Europe, 1492–1800," in *First Images of America: The Impact of the New World on the Old*, ed. Fredi Chiappelli (Berkeley: University of California Press, 1976), 815. Victoria Dickenson has applied this insight regularly in the Canadian context. See Dickenson, *Drawn from Life*; idem, "Cartier, Champlain, and the Fruits of the New World," *Scientia Canadensis* 31, no. 1–2 (2008): 27–47.

15. The distinct region that the geographer D. W. Meinig referred to as a "northern America" has often held an ambiguous place within this scholarship, yet the distinct environmental history of this region had an enormous influence on the evolution of New France. Environmental historians of these regions follow historical geographers such as Harold Innis, D. W. Meinig, and Carl O. Sauer, who were particularly attentive to the unique history of these areas. See Innis, *The Fur Trade in Canada*; D. W. Meinig, *The Shaping of America: A Geographical Perspective on 500 Years of History* (New Haven, Conn.: Yale University Press, 1986); Carl Ortwin Sauer, *Sixteenth Century North America: The Land and the People as Seen by Europeans* (Berkeley: University of California Press, 1971). Foundational environmental histories by scholars such as Bill Cronon and Carolyn Merchant focused on the region, but these were not substantially expanded upon until Joyce Chaplin's *Subject Matter*. More recently, historians such as Katherine Grandjean, Sam White, Thomas Wickman, and Anya Zilberstein have returned the study of the Northeast to a major focus of environmental historians. Chaplin, *Subject Matter*; William Cronon, *Changes in the Land: Indians, Colonists, and the Ecology of New England* (New York: Hill and Wang, 2003); Katherine A. Grandjean, "New World Tempests: Environment, Scarcity, and the Coming of the Pequot War," *William and Mary Quarterly* 68, no. 1 (2011): 75–100; Carolyn Merchant, *Ecological Revolutions: Nature, Gender, and Science in New England* (Chapel Hill: University of North Carolina Press, 1989); Sam White, "'Shewing the difference betweene their conjuration, and our invocation on the name of God for rayne': Weather, Prayer, and Magic in Early American Encounters," *William and Mary Quarterly* 72, no. 1 (2015): 33–56; Thomas Wickman, "'Winters Embittered with Hardships': Severe Cold, Wabanaki Power, and English Adjustments, 1690–1710," *William and Mary Quarterly* 72, no. 1 (2015): 57–98; Anya Zilberstein, *A Temperate Empire: Making Climate Change in Early America* (Oxford: Oxford University Press, 2016).

16. For "cultural baggage," see John Huxtable Elliott, *Empires of the Atlantic World: Britain and Spain in America, 1492–1830* (New Haven, Conn.: Yale University Press, 2006), xiii. See also Pagden, *European Encounters with the New World*, ch. 1.

17. John Huxtable Elliott, "Renaissance Europe and America: A Blunted Impact?" in *First Images of America: The Impact of the New World on the Old*, ed. Fredi Chiappelli (Berkeley: University of California Press, 1976), 11. See also idem, "Final Reflections: The Old World and the New Revisited," in *America in European Consciousness, 1493–1750*, ed. Karen Ordahl Kupperman (Chapel Hill: University of North Carolina Press, 1995), 391–408.

18. Stephen Greenblatt, *Marvelous Possessions: The Wonder of the New World* (Chicago: University of Chicago Press, 1991), 7. More broadly, this scholarship speaks to skepticism of historical accounts and our own ability to read the world that produced them. See, for example, Rita Felski, "Suspicious Minds," *Poetics Today* 32, no. 2 (2011): 215–34.

19. Alan Taylor, *American Colonies*, 24; Cronon, *Changes in the Land*, 9; Matthew Dennis, "Cultures of Nature to ca. 1810," in *A Companion to American Environmental History*, ed. Douglas Cazaux Sackman (Chichester: Wiley-Blackwell, 2010), 216.

20. In narratives of disastrous early colonial settlement, failures to properly identify new plants, animals, or peoples often foreshadow subsequent colonial failures, and the histories of the death and suffering that resulted from original misunderstandings testify to the limits of European knowledge in the first age of imperial expansion. See, for example, Dennis B. Blanton, "The Weather Is Fine, Wish You Were Here, Because I'm the Last One Alive: 'Learning' the Environment in the English New World Colonies," in *Colonization of Unfamiliar Landscapes: The Archaeology of Adaptation*, ed. Marcy Rockman and James Steele (London: Routledge, 2003), 190–200. See also Karen Ordahl Kupperman, "The Puzzle of the American Climate in the Early Colonial Period," *American Historical Review* 87, no. 5 (1982): 1262–89.

21. Sagard, *Le Grand Voyage*, 173, 311. Analyses that demonstrate how European "writing produced 'the rest of the world'" and that sensibly guide their readers toward a consideration of the specific ideological and epistemological constraints that have governed the production of colonial texts and that made colonial places knowable fit French colonial descriptions of American environments only awkwardly. Mary Louise Pratt, *Imperial Eyes: Travel Writing and Transculturation* (London: Routledge, 1992), 4.

22. Michel de Certeau, *The Writing of History*, trans. Tom Conley (New York: Columbia University Press, 1988), xxv.

23. Historian James Axtell, for example, reminds us that "the first Europeans signaled their imperial intentions by naming or renaming everything in sight" and that creating analogies was itself a political act that defied an obvious ecological difference. James Axtell, *Beyond 1492: Encounters in Colonial North America* (Oxford: Oxford University Press, 1992), 59. See also Keith Pluymers, "Taming the Wilderness in Sixteenth- and Seventeenth-Century Ireland and Virginia," *Environmental History* 16, no. 4 (2011): 617; Schiebinger, *Plants and Empire*, ch. 5.

24. Champlain, *Works*, 1:93.

25. Fischer, *Champlain's Dream*, 74–101.

26. Jacques Cartier, *The Voyages of Jacques Cartier*, ed. Ramsay Cook (Toronto: University of Toronto Press, 1993), 3, 10.

27. Dickenson, *Drawn from Life*, 22; Sauer, "Changing Perception," 815.

28. Hong Qian, "Floristic Analysis of Vascular Plant Genera of North America North of Mexico: Spatial Patterning of Phytogeography," *Journal of Biogeography* 28, no. 4 (2001): 1312.

29. Champlain, *Works*, 1:145.

30. Sagard, *Le Grand Voyage*, 311.

31. For an introduction to biogeographic thought in the work of figures such Carolus Linnaeus, Charles Darwin, and Asa Gray, see Mark V. Lomolino, Dov F. Sax, and James H. Brown, eds., *Foundations of Biogeography: Classic Papers with Commentaries* (Chicago: University of Chicago Press, 2004).

32. Jun Wen, "Evolution of Eastern Asian and Eastern North American Disjunct Distributions in Flowering Plants," *Annual Review of Ecology and Systematics* 30, no. 1 (1999): 421–55.

33. Dieter H. Mai, "Development and Regional Differentiation of the European Vegetation During the Tertiary," *Plant Systematics and Evolution* 162, no. 1–4 (1989): 83–84.

34. Ibid., 83; Qian, "Floristic Analysis," 1317; Jason D. Fridley, "Plant Invasions Across the Northern Hemisphere: A Deep-Time Perspective," *Annals of the New York Academy of Sciences* 1293, no. 1 (2013): 12.

35. Richard Ian Milne, "Northern Hemisphere Plant Disjunctions: A Window on Tertiary Land Bridges and Climate Change?" *Annals of Botany* 98, no. 3 (2006): 468.

36. Bruce H. Tiffney, "The Eocene North Atlantic Land Bridge: Its Importance in Tertiary and Modern Phytogeography of the Northern Hemisphere," *Journal of the Arnold Arboretum* 66 (1985): 243–73. Owing to seasonal variations in light exposure at different latitudes, many of the deciduous plants would likely have crossed from the west, while coniferous trees were more likely to have crossed from the east. See idem, "Geographic and Climatic Influences on the Cretaceous and Tertiary History of Euramerican Floristic Similarity," *Acta Universitatis Carolinae—Geologica* 44, no. 1 (2000): 7.

37. Tiffney, "Geographic and Climatic," 8–9.

38. Richard I. Milne and Richard J. Abbott, "The Origin and Evolution of Tertiary Relict Floras," *Advances in Botanical Research* 38 (2002): 281–314; James Cunningham Ritchie, *Post-Glacial Vegetation of Canada* (Cambridge: Cambridge University Press, 2004), 131.

39. Milne suggests that these ended 25 to 15 million years ago. Milne, "Northern Hemisphere," 466. See also Tiffney, "Geographic and Climatic," 5; Wen, "Evolution," 423.

40. The possibility of continued connections via long-distance dispersal has not been ruled out but has been judged to be unlikely. See Milne, "Northern Hemisphere." Milne and Abbott highlight that animal populations diverged before plants, suggesting continued communication of plants after that between animals had stopped. Milne and Abbott, "The Origin and Evolution of Tertiary Relict Floras," 298; Michael J. Donoghue and Stephen A. Smith, "Patterns in the Assembly of Temperate Forests Around the Northern Hemisphere," *Philosophical Transactions: Biological Sciences* 359, no. 1450 (2004): 1640.

41. Tiffney, "The Eocene"; Tiffney, "Geographic and Climatic," 6. Biogeographers nonetheless emphasize the likelihood of multiple origins and subsequent divergence. See, for example, Wen, "Evolution," 436; Qiu-Yun Xiang and Douglas E. Soltis, "Dispersal-Vicariance Analyses of Intercontinental Disjuncts: Historical Biogeographical Implications for Angiosperms in the Northern Hemisphere," *International Journal of Plant Sciences* 162, no. S6 (2001): S30. For the 10–40 million year estimate, see Milne and Abbott, "The Origin and Evolution of Tertiary Relict Floras," 299.

42. See Donald A. Levin, *The Origin, Expansion, and Demise of Plant Species* (Oxford: Oxford University Press, 2000); John N. Thompson, *The Geographic Mosaic of Coevolution* (Chicago: University of Chicago Press, 2005).

43. Brazeau, *Writing a New France*, 24–26.

44. Champlain, *Works*, 1:96.

45. Ibid.

46. Ibid., 1:22.

47. On the Little Ice Age generally, see Brian Fagan, *The Little Ice Age: How Climate Made History, 1300–1850* (New York: Basic Books, 2000).

48. Karen Ordahl Kupperman, *The Jamestown Project* (Cambridge, Mass.: Harvard University Press, 2009), 163–76.

49. Christopher J. Bilodeau, "The Paradox of Sagadahoc: The Popham Colony, 1607–1608," *Early American Studies: An Interdisciplinary Journal* 12, no. 1 (2014): 1–35.

50. Marcy Rockman, "New World with a New Sky: Climatic Variability, Environmental Expectations, and the Historical Period Colonization of Eastern North America," *Historical Archaeology* 44, no. 3 (2010): 6; David W. Stahle et al., "The Lost Colony and Jamestown Droughts," *Science* 280, no. 5363 (1998): 564–67.

51. Champlain, *Works*, 1:301. These sorts of miscalculations of colonial climates were a regular occurrence. See Kupperman, "The Puzzle of the American Climate in the Early Colonial Period"; Rockman, "New World with a New Sky," 4–20.

52. Champlain, *Works*, 1:307.

53. Ibid., 1:367; Lescarbot, *Histoire de la Nouvelle-France*, 495.

54. Champlain, *Works*, 1:367.

55. Ibid.

56. Ibid., 1:376.

57. Marie-Christine Pioffet, introduction to Marc Lescarbot, *Voyages en Acadie, 1604–1607: Suivis de la description des mœurs souriquoises comparées à celles d'autres peuples*, ed. Marie-Christine Pioffet (Sainte-Foy: Presses de l'Université Laval, 2007), 19. Climate was particularly important in discussing differences between human populations in this period. See Jan Golinski, "American Climate and the Civilization of Nature," in *Science and Empire in the Atlantic World*, ed. James Delbourgo and Nicholas Dew (London: Routledge, 2008), 153–74; Kupperman, "The Puzzle of the American Climate in the Early Colonial Period."

58. Gilbert Chinard, *L'Amérique et le rêve exotique dans la littérature française au XVIIe et au XVIIIe siècle* (Paris: Librairie Hachette 1913), 130. It was most famously José de Acosta whose passage across the "torrid zone" directly challenged his faith in classical authority. José de Acosta, *Natural and Moral History of the Indies*, trans. Frances López-Morillas (Durham, N.C.: Duke University Press, 2002), 37–39. See also Neil Safier, "The Tenacious Travels of the Torrid Zone and the Global Dimensions of Geographical Knowledge in the Eighteenth Century," *Journal of Early Modern History* 18, no. 1–2 (2014): 141–72. Latitude was equally important as a legal device for designating ownership in grants and commissions. Lauren Benton, *A Search for Sovereignty: Law and Geography in European Empires, 1400–1900* (Cambridge: Cambridge University Press, 2010); Helen Dewar, "'Y establir nostre auctorité': Assertions of Imperial Sovereignty Through Proprietorships and Chartered Companies in New France, 1598–1663" (PhD diss., University of Toronto, 2012), 33, 56.

59. *JR*, 3:47.

60. Ibid., 3:33. See also Brazeau, *Writing a New France*, 24.

61. Sagard, *Le Grand Voyage*, 147.

62. For a comprehensive overview of the evolution of fifteenth- and sixteenth-century conceptions of the world as divided into climatic zones, see Nicolás Wey Gómez, *The Tropics of Empire: Why Columbus Sailed South to the Indies* (Cambridge, Mass.: MIT Press, 2008), ch. 1. Surekha Davies argues that cartographic representations of fixed climatic bands nonetheless were fundamentally an effort to visualize the mutability of difference. Surekha Davies, *Renaissance Ethnography and the Invention of the Human: New Worlds, Maps and Monsters* (Cambridge: Cambridge University Press, 2016), 27.

63. Nicolas Denys, *Histoire naturelle des peuples, des animaux, des arbres & plantes de l'Amérique septentrionale* (Paris: Chez Claude Barbin, 1672), 5–12.

64. Brazeau, *Writing a New France*, 25–28.

65. Mary Fuller, "The Poetics of a Cold Climate," *Terrae Incognitae* 30, no. 1 (1998): 41. See also Chaplin, *Subject Matter*, ch. 2; Frank Lestringant, "Europe et théorie des climats dans la seconde moitié du XVIe siècle," in *Écrire le monde à la Renaissance: Quinze études sur Rabelais, Postel, Bodin et la littérature géographique* (Caen: Paradigme, 1993), 264–65.

66. John M. Headley, "The Sixteenth-Century Venetian Celebration of the Earth's Total Habitability: The Issue of the Fully Habitable World for Renaissance Europe," *Journal of World History* 8, no. 1 (1997): 16.

67. On sixteenth-century fisheries, see Peter E. Pope, *Fish into Wine: The Newfoundland Plantation in the Seventeenth Century* (Chapel Hill: University of North Carolina Press, 2004), ch. 1; Laurier Turgeon, "Le temps de pêches lointaines: Permanences et transformations (vers

1500–vers 1850)," in *Histoire des pêches maritimes en France*, ed. Michel Mollat (Toulouse: Privat, 1987), 134–81. For an example of the search for minerals in the arctic, see Robert McGhee, *Arctic Voyages of Martin Frobisher: An Elizabethan Adventure* (Kingston: McGill-Queen's University Press, 2001), 89–96.

68. Chaplin, *Subject Matter*, 45; Peter C. Mancall, *Hakluyt's Promise: An Elizabethan's Obsession for an English America* (New Haven, Conn.: Yale University Press, 2007), 54, 99, 149; idem, *Fatal Journey: The Final Expedition of Henry Hudson* (New York: Basic Books, 2009), 83.

69. Pioffet, introduction, 23; Marc Lescarbot, *Les muses de la Nouvelle France* (Paris: Chez A. Perier, 1618), 38–40.

70. *JR*, 3:47.

71. Ibid. Biard echoed calls for close empirical examination of American environments present throughout this early period. On the importance of empirical knowledge in the early modern Atlantic world, see Barrera-Osorio, *Experiencing Nature*; Chaplin, *Subject Matter*; Jim Egan, *Authorizing Experience: Refigurations of the Body Politic in Seventeenth-Century New England Writing* (Princeton, N.J.: Princeton University Press, 1999); Gómez, *The Experiential Caribbean*.

72. A complete history of this expansion is beyond the scope of this chapter. See Gilles Havard and Cécile Vidal, *Histoire de l'Amérique française* (Paris: Flammarion, 2004).

73. Marie-Christine Pioffet, "La Nouvelle-France dans l'imaginaire jésuite: *Terra doloris* ou Jérusalem céleste?" in *Jesuit Accounts of the Colonial Americas: Intercultural Transfers Intellectual Disputes, and Textualities*, ed. Marc André Bernier, Clorinda Donato, and Hans-Jürgen Lüsebrink (Toronto: University of Toronto Press, 2014), 329, 337–38. See also Sophie-Laurence Lamontagne, *L'hiver dans la culture québécoise, XVIIe–XIXe siècles* (Québec: Institut québécois de recherche sur la culture, 1983).

74. Pierre Boucher, *Histoire veritable et naturelle des moeurs et productions du pays de la Nouvelle France, vulgairement dite le Canada* (Boucherville: Société historique de Boucherville, 1964), 141.

75. This process is described in Lamontagne, *L'hiver dans la culture québécoise*, 39–46.

76. Kupperman, "The Puzzle of the American Climate in the Early Colonial Period."

77. Cartier, *Voyages*, 72–73.

78. Champlain, *Works*, 2:17, 60.

79. Louise Filion, Martin Lavoie, and Lydia Querrec, "The Natural Environment of the Québec City Region During the Holocene," *Post-Medieval Archaeology* 43, no. 1 (2009): 13–29; Marcel Moussette, "Between Land and Water: Being a Settler at Ile-aux-Oies in the Seventeenth and Eighteenth Centuries," *International Journal of Historical Archaeology* 12, no. 4 (2008): 279; Lydia Querrec, Louis Filion, and Réginald Auger, "Pre-European Settlement Paleoenvironments Along the Lower Saint Charles River, Québec City (Canada)," *Écoscience* 20, no. 1 (2013): 79.

80. Richard J. Huggett, *Fundamentals of Biogeography*, 2nd ed. (London: Routledge, 2004), 102.

81. Marie-Victorin calls the Laurentian forest a "penetration" of the temperate forest into an otherwise inhospitable region. Marie-Victorin, *Flore laurentienne*, 32.

82. For a discussion of forest response to climate change during the Holocene (the last 12,000 years), see Denise C. Gaudreau, "The Distribution of Late Quaternary Forest Regions in the Northeast: Pollen Data, Physiography, and the Prehistoric Record," in *Holocene Human Ecology in Northeastern North America*, ed. George P. Nicholas (New York: Plenum Press, 1988), 215–56.

83. James H. Brown and Mark V. Lomolino, *Biogeography*, 2nd ed. (Sunderland, Mass.: Sinauer Associates, 1998), 104. See also Robert Michael Morrissey, "The Power of the Ecotone: Bison, Slavery, and the Rise and Fall of the Grand Village of the Kaskaskia," *Journal of American History* 102, no. 3 (2015): 667–92.

84. Champlain, *Works*, 2:176.

85. Charles Heiser, "On Possible Sources of the Tobacco of Prehistoric Eastern North America," *Current Anthropology* 33, no. 1 (1992): 54; William Albert Setchell, "Aboriginal Tobaccos," *American Anthropologist* 23, no. 4 (1921): 403.

86. See J. S. Clark and P. D. Royall, "Transformation of a Northern Hardwood Forest by Aboriginal (Iroquois) Fire: Charcoal Evidence from Crawford Lake, Ontario, Canada," *The Holocene* 5, no. 1 (1995): 1–9; M. Day Gordon, "The Indian as an Ecological Factor in the Northeastern Forest,? *Ecology* 34, no. 2 (1953): 329–46; Joseph D. Mitchell, "The American Indian: A Fire Ecologist," *American Indian Culture and Research Journal* 2, no. 2 (1978): 26–31; Nancy J. Turner, Iain J. Davidson-Hunt, and Michael O'Flaherty, "Living on the Edge: Ecological and Cultural Edges as Sources of Diversity for Social-Ecological Resilience," *Human Ecology* 31, no. 3 (2003): 439–61.

87. Denis Jamet, "Relation du Père Denis Jamet, Récollet de Québec, au cardinal de Joyeuse," in *Nouveaux documents sur Champlain et son époque, Volume 1 (1560–1622)*, ed. Robert Le Blant and René Baudry (Ottawa: Public Archives of Canada, 1967), 349.

88. See Jacques Rousseau, "Pour une esquisse biogéographique du Saint-Laurent," *Cahiers de géographie du Québec* 11, no. 23 (1967): 181–214.

89. Victoria Dickenson, "The Herons Are Still Here: History and Place," in *Metropolitan Natures: Environmental Histories of Montréal*, ed. Stéphane Castonguay and Michèle Dagenais (Pittsburgh: University of Pittsburgh Press, 2011), 45–46. See also Conrad Heidenreich, *Explorations and Mapping of Samuel de Champlain, 1603–1632* (Toronto: University of Toronto Press, 1976).

90. Marie-Victorin, *Flore laurentienne*, 23. On the relationship between climate patterns and aboriginal settlement, see Robert J. Hasenstab, "Aborignal Settlement Patterns in Late Woodland Upper New York State," in *Archaeology of the Iroquois: Selected Readings and Research Sources*, ed. Jordan E. Kerber (Syracuse, N.Y.: Syracuse University Press, 2007), 164–73.

91. Champlain, *Works*, 1:144.

92. Ibid., 1:376.

93. *JR*, 9:211–13.

94. On the particularly harsh period of the seventeenth century, see Geoffrey Parker, *Global Crisis: War, Climate Change and Catastrophe in the Seventeenth Century* (New Haven, Conn.: Yale University Press, 2014), ch. 1. Knowledge of the climate in New France increased dramatically in the seventeenth century. See Colin Coates and Dagomar Degroot, "'Les bois engendrent les frimas et les gelées': Comprendre le climat en Nouvelle-France," *Revue d'histoire de l'Amérique française* 68, no. 3–4 (2015): 197–219; Yvon Desloges, *Sous les cieux de Québec: Météo et climat, 1534–1831* (Québec: Septentrion, 2016).

95. *JR*, 3:63.

96. Champlain, *Works*, 2:17. On Champlain's geographical discourse, see Christian Morissonneau, *Le langage géographique de Cartier et de Champlain: Choronymie, vocabulaire et perception* (Québec: Presses de l'Université Laval, 1978), 216–22.

97. Champlain, *Works*, 2:20.

98. Ibid., 2:22.

99. Sagard, *Le Grand Voyage*, 136.

100. Karen Ordahl Kupperman, "Fear of Hot Climates in the Anglo-American Colonial Experience," *William and Mary Quarterly* 41, no. 2 (1984): 219.

101. Novel plants seem to have featured more prominently in visual representations than they did in written accounts. See, for an analysis of plants depicted in Champlain's maps, Pierre-Simon Doyon, "L'iconographie botanique en Amérique française aux 17e et 18e siècles," http://www.uqtr.ca/arts.histoire.botanique.

102. Champlain, *Works*, 3:50–51.

103. Sagard, *Le Grand Voyage*, 312.

104. Louis Nicolas, *The Codex Canadensis and the Writings of Louis Nicolas*, ed. François-Marc Gagnon, trans. Nancy Senior and Réal Ouellet (Kingston: McGill-Queen's University Press, 2011), 283. For closer study of Nicolas's botanical names, see Doyon, "L'iconographie botanique"; Marthe Faribault, "Les phytonymes de l'*Histoire naturelle des Indes occidentales* de Louis Nicolas: Image du lexique botanique canadien à la fin du XVIIe siècle," in *Français du Canada, Français de France: Actes du quatrième colloque international de Chicoutimi, Québec, du 21–24 septembre 1994*, ed. Thomas Lavoie (Tübingen: Max Niemayer Verlag, 1996), 99–113. A recent edition of Nicolas's writings also demonstrates that the Jesuit's work was part of a larger episteme that sought out similarity. François-Marc Gagnon, "Louis Nicolas's Depiction of the New World in Figures and Text," in *Codex Canadensis*, 30.

105. *JR*, 51:121.

106. Lescarbot, *Histoire de la Nouvelle-France*, 536.

107. Champlain, *Works*, 2:10.

108. This section is influenced by ethnobotanical and ethnobiological works such as Scott Atran, *Cognitive Foundations of Natural History: Towards an Anthropology of Science* (Cambridge: Cambridge University Press, 1992); Brent Berlin, *Ethnobiological Classification: Principles of Categorization of Plants and Animals in Traditional Societies* (Princeton, N.J.: Princeton University Press, 1992).

109. Berlin, *Ethnobiological Classification*, 76.

110. Champlain, *Works*, 2:341.

111. Nicolas, *Codex Canadensis*, 281.

112. Ibid., 280. These descriptions therefore exhibited a tendency for authors to describe American natural objects by "decontextualizing them and implicitly denying that their native habitat or setting mattered." Karen Ordahl Kupperman, "Introduction: The Changing Definition of America," in *America in European Consciousness, 1493–1750*, ed. Karen Ordahl Kupperman (Chapel Hill: University of North Carolina Press, 1995), 12.

113. The Columbian Exchange was the transatlantic circulation of plants, animals, and diseases in the wake of European voyages of exploration and colonization. Alfred W. Crosby, *The Columbian Exchange: Biological and Cultural Consequences of 1492* (Westport, Conn.: Greenwood Press, 1972).

114. Michèle Bilimoff, ed., *Promenades dans des jardins disparus: D'après les Grandes Heures d'Anne de Bretagne* (Rennes: Ouest-France, 2005), 92; Harry S. Paris et al., "First Known Image of Cucurbita in Europe, 1503–1508," *Annals of Botany* 98 (2006): 41–47.

115. Jules Janick and Harry S. Paris, "The Cucurbit Images (1515–1518) of the Villa Farnesina, Rome," *Annals of Botany* 97 (2006): 165–76; Paris et al., "First Known Image," 41.

116. Paris et al., "First Known Image," 46.

117. Ibid.

118. Charles Estienne and Jean Liebault, *L'agriculture et maison rustique* (Paris, 1586), 302.

119. Ibid., 303.

120. Pedro Revilla Temiño et al., "Isozyme Variability Among European Maize Populations and the Introduction of Maize in Europe," *Maydica* 48, no. 2 (2003): 141–52. See also C. Mir et al., "Out of America: Tracing the Genetic Footprints of the Global Diffusion of Maize," *Theoretical and Applied Genetics* 126 (2013): 2671–82.

121. Pierre Ponsot, "Les débuts du maïs en Bresse sous Henri IV," *Histoires & Sociétés Rurales* 23, no. 1 (2005): 117–36. This is also when Europeans began to first take note of large-scale cultivation of corn in Africa. James C. McCann, *Maize and Grace: Africa's Encounter with a New World Crop* (Cambridge, Mass.: Harvard University Press, 2005).

122. Estienne and Liebault, *Maison rustique*, 106–7. See also Lien Speleers and Jan M. A. dan der Valk, "Economic Plants from Medieval and Post-Medieval Brussels (Belgium), an Overview of the Archaeobotanical Records," *Quaternary International* 436 (2017): 96–109.

123. Charles Estienne and Jean Liebault, *L'agriculture et maison rustique* (Paris, 1640), 217. For changing French perceptions of tobacco in the seventeenth century, see Christopher M. Parsons, "Of Natives, Newcomers, and Nicotiana: Tobacco in the History of the Great Lakes Region," in *French and Indians in the Heart of North America, 1630–1815*, ed. Robert Englebert and Guillaume Teasdale (East Lansing: Michigan State University Press, 2013), 21–42.

124. Nicolas, *Codex Canadensis*, 291.

125. Sagard, *Le Grand Voyage*, 312.

126. Lescarbot, *Voyages en Acadie*, 113.

127. Cartier, *Voyages*, 80. On the tree's Atlantic history, and the problems with offering a botanical identification, see Jacques Mathieu, *L'Annedda: L'arbre de vie* (Québec: Septentrion, 2009).

128. Champlain, *Works*, 1:322.

129. Dickenson, *Drawn from Life*, 74–75.

130. Georges E. Sioui, *Huron-Wendat: The Heritage of the Circle*, trans. Jane Brierley (Vancouver: University of British Columbia Press, 1999), 220.

131. Sagard, *Le Grand Voyage*, 311.

132. *JR*, 43:147.

133. Nicolas, *Codex Canadensis*, 417.

134. Indigenous names and knowledge could become standardized and appropriated as they increasingly circulated in texts. See Daniela Bleichmar, "Books, Bodies, and Fields: Sixteenth-Century Transatlantic Encounters with New World Materia Medica," in *Colonial Botany: Science, Commerce, and Politics in the Early Modern World*, ed. Claudia Swan and Londa Schiebinger (Philadelphia: University of Pennsylvania Press, 2007), 91–95.

135. Gédéon de Catalogne, "Mémoire de Gédéon de Catalogne sur les plans des seigneuries et habitations des gouvernements de Québec, les Trois-Rivières et Montréal," *Bulletin des recherches historiques* 21, no. 9 (1915): 262, cited at "Trésor de la langue française au Québec," http://www.tlfq.ulaval.ca/.

136. Ibid.

137. Andrew Warnes, *Savage Barbecue: Race, Culture, and the Invention of America's First Food* (Athens: University of Georgia Press, 2008).

138. From the first voyages of Jacques Cartier in the sixteenth century, the presence of grapes was read as a symbol that French culture could survive in New France. See Brazeau, *Writing a New France*, ch. 2; Ferland, *Bacchus en Canada*, 28–30.

139. Champlain, *Works*, 1:129; References like this are countless. See, for example, René Cuillerier, *Nation Iroquoise: A Seventeenth-Century Ethnography of the Iroquois*, ed. José António Brandão (Lincoln: University of Nebraska Press, 2003), 52; Denys, *Histoire naturelle des*

peuples, 51; Louis Hennepin, *Description de la Louisiane nouvellement découverte au sud oüest de la Nouvelle France* (Paris: Chez la veuve Sébastien Huré, 1683), 2.

140. *JR*, 51:121.

141. The filtering of substantial and accidental traits was of broader intellectual interest to renaissance and early modern naturalists. See Atran, *Cognitive Foundations of Natural History*, 153–54.

142. Dickason, *The Myth of the Savage*, 63.

143. This sixteenth-century legacy was produced in dialogue with the colonial efforts of French Protestants. See Lestringant, *Le Huguenot et le sauvage*.

144. Léon Pouliot, "Premières pages du journal des Jésuites de Québec, 1632–1645," *Rapport de l'Archiviste de la Province de Québec* 41 (1963): 115.

145. Chinard, *L'Amérique et le rêve exotique*, v.

146. Michel de Montaigne, *Oeuvres completes*, ed. Albert Thibaudet and Maurice Rat (Paris: Gallimard, 1962), 203. See also Sankar Muthu, *Enlightenment Against Empire* (Princeton, N.J.: Princeton University Press, 2003), 15; Michael Wintroub, *A Savage Mirror: Power, Identity, and Knowledge in Early Modern France* (Stanford, Calif.: Stanford University Press, 2006), 9–10.

147. This was particularly prevalent in debates about the status of alchemy. See William R. Newman, *Promethean Ambitions: Alchemy and the Quest to Perfect Nature* (Chicago: University of Chicago Press, 2004). More broadly these engaged debates about the proper purpose of the sciences and the moral status of the services that they offered to the public and the state. See Vera Keller, *Knowledge and the Public Interest, 1575–1725* (Cambridge: University of Cambridge Press, 2015), particularly ch. 5.

148. This ambivalence has been highlighted by numerous scholars. See Chinard, *L'Amérique et le rêve exotique*, 138; Sayre, *Les Sauvages Américains*, 128; Micah True, *Masters and Students: Jesuit Mission Ethnography in Seventeenth-Century New France* (Kingston: McGill-Queen's University Press, 2015), ch. 3.

149. Henri Joutel, *Journal historique du dernier voyage que feu M. de LaSale fit dans le golfe de Mexique, pour trouver l'embouchure, & le cours de la riviere de Missicipi, nommeé à present la riviere de Saint Loüis, qui traverse la Louisiane: Où l'on voit l'histoire tragique de sa mort, & plusieurs choses curieuses du nouveau monde* (Paris: Chez E. Robinot, 1713), viii.

150. This was common practice in New England as well. Cronon, *Changes in the Land*, 90; Brian Donahue, *The Great Meadow: Farmers and the Land in Colonial Concord* (New Haven, Conn.: Yale University Press, 2004), 87–89.

151. Guimont, *La Petite-ferme du cap Tourmente*, ch. 1. For a discussion of the agricultural practices of the Saint Lawrence Iroquois, as well as a survey of the debate over their disappearance, see Roland Tremblay, *Les Iroquoiens du Saint-Laurent: Peuple du maïs* (Montréal: Editions de l'Homme, 2006).

152. Michael A. LaCombe, *Political Gastronomy: Food and Authority in the English Atlantic World* (Philadelphia: University of Pennsylvania Press, 2012), 4; Kathleen Donegan, *Seasons of Misery: Catastrophe and Colonial Settlement in Early America* (Philadelphia: University of Pennsylvania Press, 2013), 123–24.

153. This is further discussed in Chapter 3.

154. Lescarbot, *Histoire de la Nouvelle-France*, 555.

155. *JR*, 21:63, 119. Pioffet writes that missionaries often emphasized the density of the woods (and ice). Pioffet, "La Nouvelle-France dans l'imaginaire jésuite," 329.

156. Champlain, *Works*, 2:127.

157. Ibid., 2:267.

158. *JR*, 5:123.

159. Ibid., 6:161.

160. Numerous historians have also commented upon the failure to appreciate the complexity of indigenous ecological knowledge that focused on seasonal occupation of multiple sites. See, for example, Cronon, *Changes in the Land*, 54–55; Karen Ordahl Kupperman, *Indians and English: Facing Off in Early America* (Ithaca, N.Y.: Cornell University Press, 2000), 160–61; David J. Silverman, "'We Chuse to Be Bounded': Native American Animal Husbandry in Colonial New England," *William and Mary Quarterly* 60, no. 3 (2003): 511–48.

161. Champlain, *Works*, 3:38.

162. Ibid., 3:44.

163. Ibid., 3:125.

164. Ibid., 3:156.

CHAPTER 2

1. Lescarbot, *Histoire de la Nouvelle-France*, 454.

2. Ibid., 561, 600.

3. Ibid., 500.

4. Ibid., 474, 493.

5. Lescarbot, for example, described the close observation of plots that had experimentally been sown with wheat at different dates to see how the grains came to maturity. Ibid., 573.

6. Ibid., 541.

7. Joutel, *Journal historique*, viii.

8. This reimagining of agricultural labor in early modern France has been best explored in Duport, *Le jardin et la nature*, esp. ch. 1. See also Sara E. Melzer, "The Role of Culture and Art in France's Colonial Strategy of the Seventeenth Century," in *Jesuit Accounts of the Colonial Americas: Intercultural Transfers, Intellectual Disputes, and Textualities*, ed. Marc André Bernier, Clorinda Donato, and Hans-Jürgen Lüsebrink (Toronto: University of Toronto Press, 2014), 171–75.

9. Chandra Mukerji highlights the French term for this stewardship: *mesnagement*. Chandra Mukerji, "Stewardship Politics and the Control of Wild Weather: Levees, Seawalls, and State Building in 17th-Century France," *Social Studies of Science* 37, no. 1 (2007): 127–33. See also idem, *Territorial Ambitions and the Gardens of Versailles* (Cambridge: Cambridge University Press, 1997).

10. This agricultural metaphor has been discussed by numerous scholars. See, for example, Carole Blackburn, *Harvest of Souls: The Jesuit Missions and Colonialism in North America, 1632–1650* (Kingston: McGill-Queen's University Press, 2000); Melzer, *Colonizer or Colonized*, ch. 8; Meridith Beck Sayre, "Cultivating Soils and Souls: The Jesuit Garden in the Americas" (master's thesis, Simon Fraser University, 2007).

11. Champlain, *Works*, 2:29. Other founders also gathered historical information about agriculture from indigenous peoples. Colin Coates, "The Colonial Landscape of the Early Town," in *Metropolitan Natures: Environmental Histories of Montreal*, ed. Stéphane Castonguay and Michèle Dagenais (Pittsburgh: University of Pittsburgh Press, 2011), 22.

12. Champlain, *Works*, 2:179.

13. Lescarbot, *Histoire de la Nouvelle-France*, 544.

14. Ibid., 441.

15. *JR*, 9:191.

16. Ibid., 4:195.

17. Pouliot, "Premières pages du journal des Jésuites de Québec," 117.

18. A history of agriculture in French North America is beyond the scope of this present chapter. For work on orchards, see Sylvie Dépatie, "Jardins et vergers à Montréal au XVIIIe siècle," in *Vingt ans apres, Habitants et marchands: Lectures de l'histoire des XVIIe et XVIIIe siècles canadiens*, ed. Sylvie Dépatie et al. (Kingston: McGill-Queen's University Press, 1998), 226–53. For work on agriculture in New France, see Allan Greer, *Peasant, Lord and Merchant: Rural Society in Three Quebec Parishes, 1740–1840* (Toronto: University of Toronto Press, 1985); Louise Dechêne, *Habitants and Merchants in Seventeenth Century Montreal*, trans. Liana Vardi (Kingston: McGill-Queen's University Press, 1992). For work on agriculture in the *pays d'en haut* (including the Great Lakes and upper Louisiana), see Carl J. Ekberg, *French Roots in the Illinois Country: The Mississippi Frontier in Colonial Times* (Champaign: University of Illinois Press, 2000); M. J. Morgan, *Land of Big Rivers: French and Indian Illinois, 1699–1778* (Carbondale: Southern Illinois University Press, 2010); Joseph Zitomersky, *French Americans—Native Americans in Eighteenth-Century French Colonial Louisiana: The Population Geography of the Illinois Indians, 1670s–1760s: The Form and Function of French-Native Settlement Relations in Eighteenth Century Louisiana* (Lund: Lund University Press, 1994), particularly the supplementary section. For work on agriculture in Louisiana, see Daniel H. Usner, *Indians, Settlers, & Slaves in a Frontier Exchange Economy: The Lower Mississippi Valley Before 1783* (Chapel Hill: University of North Carolina Press, 1992), chs. 5 and 6.

19. Lescarbot, *Histoire de la Nouvelle-France*, 474.

20. Bernard Palissy, *Oeuvres complètes de Bernard Palissy* (Paris: J. J. Dubochet, 1844), 169.

21. On these terms and their evolution, see Kenneth Woodbridge, *Princely Gardens: The Origins and Development of the French Formal Style* (New York: Rizzoli, 1986).

22. Champlain, *Works*, 1:371–73.

23. Duport, *Le jardin et la nature*, 30–34.

24. Hunt, *Greater Perfections*, ch. 3.

25. Bronwen Catherine McShea, "Presenting the 'Poor Miserable Savage' to French Urban Elites: Commentary on North American Living Conditions in Early Jesuit Relations," *Sixteenth Century Journal* 44, no. 3 (2013): 683–711; Dominique Deslandres, *Croire et faire croire: Les missions françaises au XVIIe siècle (1600–1650)* (Paris: Fayard, 2002).

26. Lescarbot, *Histoire de la Nouvelle-France*, 599–600; Cornelius J. Jaenen, "'Les Sauvages Ameriquains': Persistence into the 18th Century of Traditional French Concepts and Constructs for Comprehending Amerindians," *Ethnohistory* 29, no. 1 (1982): 45.

27. For a study of how both England and France defined their own imperial projects in contrast to Spanish conquest, see Pagden, *Lords of All the World*.

28. Mukerji, "Stewardship Politics," 128.

29. This is most fully developed in Mukerji's analysis of the gardens at Versailles. Mukerji, *Territorial Ambitions*.

30. Ibid., 8.

31. Duport, *Le jardin et la nature*, 57.

32. Woodbridge, *Princely Gardens*, 52.

33. This work is synonymous with that of André Le Nôtre. See Thierry Mariage, *The World of André Le Nôtre*, trans. Graham Larkin (Philadelphia: University of Pennsylvania Press, 1998).

34. Duport, *Le jardin et la nature*, ch. 3; Andrew McRae, *God Speed the Plough: The Representation of Agrarian England, 1500–1660* (Cambridge: Cambridge University Press, 1996), chs. 6, 7.

35. Danièle Duport, "La 'science' d'Olivier de Serres et la connaissance du 'naturel,'" *Bulletin de l'Association d'étude sur l'humanisme, la réforme et la renaissance* 50 (2000): 88.

36. Serres, *Le théatre d'agriculture et mésnage des champs*, preface. See also Duport, *Le jardin et la nature*, 30.

37. Serres, *Le théatre d'agriculture et mésnage des champs*, n.p.

38. Mariage, *The World of André Le Nôtre*, 1.

39. For a study of the classical influence on Serres, see Martine Gorrichon, "Les travaux et les jours à Rome et dans l'ancienne France: Les agronomes latins inspirateurs d'Olivier de Serres" (PhD diss., Université de Tours, 1976).

40. Melzer, *Colonizer or Colonized*, 7.

41. Pagden, *Lords of All the World*, ch. 1.

42. Melzer, *Colonizer or Colonized*, 206–8.

43. Barbiche, "Henri IV and the World Overseas: A Decisive Time in the History of New France," in *Champlain: The Birth of French America*, ed. Raymonde Litalien and Denis Vauegeois (Montréal: McGill-Queen's University Press, 2004), 29–32.

44. Melzer, *Colonizer or Colonized*, ch. 8.

45. See, for a discussion of the dispute over royal support during this time, Dewar, "'Y establir nostre auctorité,'" ch. 1. See also Barbiche, "Henri IV and the World Overseas."

46. Lescarbot, *Histoire de la Nouvelle-France*, 604.

47. For a discussion of Lescarbot's study of novel plants, see Asselin, Cayouette, and Mathieu, *Curieuses histoires de plantes du Canada*, 1:108–15.

48. Lescarbot, *Histoire de la Nouvelle-France*, 604. This passage therefore echoes broader comparisons that the French were making to the Romans during this time. See Melzer, *Colonizer or Colonized*.

49. Lescarbot, *Histoire de la Nouvelle-France*, 557.

50. Bernard Allaire, *Pelleteries, manchons et chapeaux de castor: Les fourrures nord-américaines à Paris, 1500–1632* (Sillery: Septentrion, 1999), 91.

51. Christian Feest, "The Collecting of American Indian Artifacts in Europe, 1493–1750," in *America in European Consciousness, 1493–1750*, ed. Karen Ordahl Kupperman (Chapel Hill: University of North Carolina Press, 1995), 329; see also idem, *Premières nations, collections royales: Les Indiens des forêts et des prairies d'Amérique du Nord* (Paris: Musée du Quai Branly, 2007).

52. As Anthony Alan Shelton explains, "The cultural origins of these items seem to have been of less importance than their broad geographical provenance." Anthony Alan Shelton, "Cabinets of Transgression: Renaissance Collections and the New World," in *The Cultures of Collecting*, ed. John Elsner and Roger Cardinal (London: Reaktion Books, 1994), 184.

53. On this "culture of curiosity," see Krzysztof Pomian, *Collectionneurs, amateurs et curieux: Paris, Venise, XVIe–XVIIIe siècle* (Paris: Gallimard, 1987), 61–80.

54. This is best analyzed in Cooper, *Inventing the Indigenous*. See also Benjamin Schmidt, *Inventing Exoticism: Geography, Globalism, and Europe's Early Modern World* (Philadelphia: University of Pennsylvania Press, 2015).

55. This is discussed more in Chapter 4.

56. Letter to Baron d'Alegre, 1630, in Philippe Tamizey de Larroque, ed., *Lettres de Peiresc* (Paris: Imprimerie nationale, 1898), 7:21–22. This letter is also discussed in Dickenson, "Cartier, Champlain, and the Fruits of the New World," 44.

57. Peiresc was part of an extended network of collectors. Dickenson, "Cartier, Champlain, and the Fruits of the New World," 43.

58. Maurice Bouvet, "Les anciens jardins botaniques médicaux de Paris," *Revue d'histoire de la pharmacie* 35, no. 199 (1947): 226–28; Yves Laissus, "Le Jardin du roi," in *Enseignement et diffusion des sciences en France au XVIIIe siècle,* ed. René Taton (Paris: Hermann, 1964), 288–92.

59. Elizabeth Hyde, "Flowers of Distinction: Taste, Class, and Floriculture in Seventeenth-Century France," in *Bourgeois and Aristocratic Encounters in the Garden,* ed. Michael Conan (Washington, D.C.: Dumbarton Oaks Research Library and Collection, 2002), 80.

60. Jacques Mathieu, *Le premier livre de plantes du Canada: Les enfants des bois du Canada au jardin du roi à Paris en 1635* (Sainte-Foy: Presses de l'Université Laval, 1998), 69; C. Laflamme, "Jacques-Philippe Cornuti: Note pour servir à l'histoire des sciences au Canada," *Mémoires et comptes rendus de la Société Royale du Canada,* 2nd ser., 8, no. 4 (1901): 4, 61.

61. Mathieu, *Le premier livre,* 69–71.

62. Ibid., 71.

63. Guy de La Brosse, *Catalogue des plantes cultivées a présent au Jardin royal des plantes medicinales, establi par Louis Le Juste, à Paris* (Paris, 1641), n.p.

64. See, for a description of this world, Bouvet, "Les anciens jardins."

65. Mathieu, *Le premier livre de plantes,* 37–38.

66. See, for example, the place of Robin in the networks of Carolus Clusius, discussed in Florike Egmond, *World of Carolus Clusius: Natural History in the Making, 1550–1610* (London: Pickering and Chatto Publishers, 2010), 133–38.

67. Mathieu, *Le premier livre,* 37–38.

68. This claim has remained the subject of some debate. See Frederick J. Peabody, "A 350-Year-Old American Legume in Paris," *Castanea* 47, no. 1 (1982): 99–104.

69. Mathieu, *Le premier livre,* 71.

70. Pierre-Joseph Buc'hoz, *Dictionnaire universel des plantes, arbres et arbustes de la France* (Paris: Lacombe, 1770–71), 1:16–17. The tree is also described in Nicolas Lemery, *Traité universel des drogues simples mises en ordre alphabétique* (Paris: L. d'Houry, 1698), 625. The tree quickly also spread throughout Europe. Michaela Vítková et al., "Black Locust (*Robinia Pseudoacacia*) Beloved and Despised: A Story of an Invasive Tree in Central Europe," *Forest Ecology and Management* 384 (2016): 287–302.

71. The plant is given its modern name in Carl von Linné, *Species plantarum exhibentes plantas rite cognitas, ad genera relatas: Cum differentiis specificis, nominibus trivialibus, synonymis selectis, locis natalibus, secundum systema sexuale digestas* (Stockholm: Laurentii Salvii, 1753), 2:722. Already by the early seventeenth century, botanists recognized that it was not a true acacia. See John Parkinson, *Theatrum Botanicum: The Theater of Plants* (London: Thomas Cotes, 1640), 1550.

72. Peabody, "A 350-Year-Old American Legume in Paris," 100; Margaret Willes, *Making of the English Gardener: Plants, Books and Inspiration, 1550–1660* (New Haven, Conn.: Yale University Press, 2011), 148.

73. Peabody, "A 350-Year-Old American Legume in Paris," 100–101. For a discussion of Champlain's botanical observations on this voyage, see Jacques Rousseau, "Samuel de Champlain, botaniste mexicain et antillais," *Cahier des dix* 16 (1951): 39–61.

74. The most recent reprint of this text is Jacques Philippe Cornut, *Canadensium plantarum aliarumque nondum editarum historia* (New York: Johnson Reprint Corp., 1966). For an analysis of the origins of Cornut's plants, see James S. Pringle, "How 'Canadian' Is Cornut's

Canadensium Plantarum Historia? A Phytogeographic and Historic Analysis," *Canadian Horticultural History: An Interdisciplinary Journal* 1, no. 4 (1988): 190–209.

75. Cooper, *Inventing the Indigenous*, 55–57; Dickenson, *Drawn from Life*, 78–81.

76. *JR*, 10:103.

77. Marie-Christine Pioffet, *La tentation de l'épopée dans les Relations des jésuites* (Sillery: Septentrion, 1997), 18.

78. *JR*, 59:103.

79. Ibid., 3:236.

80. Ibid., 3:238.

81. Ibid., 2:201.

82. Ibid., 4:111.

83. Normand Doiron, *L'art de voyager: Le déplacement à l'époque classique* (Sainte-Foy: Presses de l'Université Laval, 1995), 34; Sayre, *Les Sauvages Américains*, 84.

84. See, for example, a table that shows divergent formal evolution in Pierre Berthiaume, *L'aventure américaine au XVIIIe siècle: Du voyage à l'écriture* (Ottawa: Presses de l'Université d'Ottawa, 1990), 5.

85. Tim Ingold, *The Perception of the Environment: Essays on Livelihood, Dwelling and Skill* (London: Routledge, 2000), ch. 10.

86. Denys, *Histoire naturelle des peuples*, 325.

87. Timothy Silver, for example, has written that "not only did colonists come from Europe, but the European market frequently dictated how colonists used the land and resources of America." Timothy Silver, *A New Face on the Countryside: Indians, Colonists, and Slaves in South Atlantic Forests, 1500–1800* (Cambridge: Cambridge University Press, 1990), 4. Matthew Dennis writes that "Americans would continue to see the continent as a magazine of resources and commodities and justify its expropriation through its exploitation and transformation." Dennis, "Cultures of Nature to ca. 1810," 215. See also Cronon, *Changes in the Land*, 166–70.

88. Monique Delaney, "'Le Canada est un païs de Bois': Forest Resources and Shipbuilding in New France, 1660–1760" (PhD diss., McGill University, 2003), 132.

89. Sagard, *Le Grand Voyage*, 145.

90. Ibid.

91. For a discussion of the rise of travel literature as an authoritative genre in this period, see Mary B. Campbell, *The Witness and the Other World: Exotic European Travel Writing, 400–1600* (Ithaca, N.Y.: Cornell University Press, 1988); Joan-Pau Rubiés, "Travel Writing as a Genre: Facts, Fictions and the Invention of a Scientific Discourse in Early Modern Europe," *International Journal of Travel and Writing* 5, no. 3 (2000): 5–33.

92. This should not, however, be thought to indicate an indifference to the environment itself as some have suggested. See Chinard, *L'Amérique et le rêve exotique*, 100.

93. Brazeau, *Writing a New France*, 4–5.

94. As two recent scholars of travel literature in this period have outlined, those who have studied the *récit de voyage* maintain "an ambivalent relationship with the notion of 'genre.'" Grégoire Holtz and Vincent Masse, "Étudier les récits de voyage: Bilan, questionnements, enjeux," *Arborescences*, no. 2 (2012): 7.

95. Paolo Carile, *Le regard entravé: Littérature et anthropologie dans les premiers textes sur la Nouvelle-France* (Sillery: Septentrion, 2000), 40.

96. Gabriel Sagard, *Dictionaire de la langve Hvronne, necessaire à ceux qui n'ont l'intelligence d'icelle, & ont à traiter auec les sauuages du pays* (Paris: Librairie Tross, 1865), n.p.

97. Carla Zecher, "Marc Lescarbot Reads Jacques Cartier: Colonial History in the Service of Propaganda," *L'Esprit Créateur* 48, no. 1 (2008): 107.

98. Nicolas, *Codex Canadensis*, 27–28; Germaine Warkentin, "Aristotle in New France: Louis Nicolas and the Making of the Codex Canadensis," *French Colonial History* 11, no. 1 (2010): 98–99.

99. Boucher, *Histoire veritable*, n.p. See also Berry, "The Delights of Nature," 224; Séraphin Marion, "Pierre Boucher, Ecrivain (1927)," in *Histoire veritable et naturelle des moeurs et productions du pays de la Nouvelle France, vulgairement dite le Canada* (Boucherville: Société historique de Boucherville, 1964), 236–47.

100. Nicolas Denys, *Description geographique et historique des costes de l'Ameriqve Septentrionale avec l'Histoire naturelle du païs* (Paris: L. Billaine, 1672), n.p.

101. See, for overviews of this evolution, Cooper, *Inventing the Indigenous*; Ogilvie, *The Science of Describing*. One essay argues that during this period "it would be more correct to talk of a proliferation of natural histories—in the plural—all with different philosophical pedigrees and correspondingly different notions of what historia was about." Gianna Pomata and Nancy G. Siraisi, introduction to *Historia: Empiricism and Erudition in Early Modern Europe*, ed. Gianna Pomata and Nancy G. Siraisi (Cambridge, Mass.: MIT Press, 2005), 2.

102. See the "Table des chapitres" in Boucher, *Histoire veritable*.

103. Atran, *Cognitive Foundations of Natural History*, 30–31.

104. Doyon, "L'iconographie botanique."

105. This was also the case in seventeenth-century England. See Barbara J. Shapiro, *A Culture of Fact: England, 1550–1720* (Ithaca, N.Y.: Cornell University Press, 2003), 70.

106. Hennepin, *Description*, 51–52.

107. Ibid.

108. As several commentators have demonstrated, Lescarbot also worked closely with classical and historical sources. See Frank Lestringant, "Champlain, Lescarbot et la 'conférence' des histoires," in *L'expérience huguenote au Nouveau Monde (XVIe siècle)* (Geneva: Librairie Droz, 1996), 329–44; Marie-Christine Pioffet, "Marc Lescarbot sur les traces de Pline l'Ancien," *Renaissance and Reformation* 24, no. 3 (2000): 5–15; Zecher, "Marc Lescarbot."

109. Jack Warwick, introduction to *Le Grand Voyage du pays des Hurons: Suivi du Dictionnaire de la langue huronne*, ed. Jack Warwick (Montréal: Presses de l'Université de Montréal, 1998), 34.

110. Quoted in Jack Warwick, "Gabriel Sagard's 'je' in the First *Histoire du Canada*," in *Reflections: Autobiography and Canadian Literature*, ed. K. P. Stich (Ottawa: University of Ottawa Press, 1988), 29.

111. This is a primary characteristic of the travel narrative in general. See Kai Mikkonen, "The 'Narrative Is Travel' Metaphor," *Narrative* 15, no. 3 (2007): 286–305. This is explicitly discussed in the context of New France in Sayre, *Les Sauvages Américains*.

112. *JR*, 3:53–55.

113. Champlain, *Works*, 1:489; Lescarbot, *Histoire de la Nouvelle-France*, 604.

114. For a study of the empirical culture of early French explorers, see François-Marc Gagnon, "Experientia est rerum magistra—Savoir empirique et culture savante chez les premiers voyageurs au Canada," *Questions de culture* 1 (1981): 47–61. For a broader discussion of the importance of the "autoptic imagination" in the discovery of the Americas, see Pagden, *European Encounters with the New World*, 51–56. See also Grafton, *New Worlds, Ancient Texts*.

115. This was particularly true when dealing with supernatural phenomena. See Pioffet, *La tentation*, ch. 7. This is a process that Doiron suggests was happening more broadly in French

literature. Doiron, *L'art de voyager*, 51–53. See also, for a study of how these texts were collected and edited, Grégoire Holtz, *L'ombre de l'auteur: Pierre Bergeron et l'écriture du voyage à la fin de la Renaissance* (Geneva: Librairie Droz, 2011); Andreas Motsch and Grégroire Holtz, eds., *Éditer la Nouvelle France* (Québec: Presses de l'Université Laval, 2011).

116. *JR*, 5:99.

117. Ibid., 51:121.

118. Literary scholar Marie-Christine Pioffet argues that Jesuit accounts both expanded the field of their own experience and added details to their accounts that they could not possibly have lived. Both, she suggests, were the products of an after-the-fact rationalization of extraordinary and otherwise unexplainable events that took place in the Society of Jesus' American missions. Pioffet ultimately contends that this "come and go between the real and the imaginary" contributed to the construction of the *Relations* as an epic history. Pioffet, *La tentation*, 187.

119. Jim Egan (*Authorizing Experience*) has explained that John Smith also embraced his experience as a source of epistemological authority.

120. Brazeau, *Writing a New France*, 30. See also, for a longer discussion of Champlain's style of authority, John A. Dickinson, "Champlain, Administrator," in *Champlain: The Birth of French America*, 211–17.

121. Champlain, *Works*, 3:205.

122. Also see the work of Thomas Worcester and Micah True, who both argue that Jesuits intended their texts to legitimate their position as authorities on American peoples and spirituality. See Thomas Worcester, "A Defensive Discourse: Jesuits on Disease in Seventeenth-Century New France," *French Colonial History* 6 (2005): 1–15; Micah True, "Maistre et Escolier: Amerindian Languages and Seventeenth-Century French Missionary Politics in the Jesuit Relations from New France," *Seventeenth-Century French Studies* 31, no. 1 (2009): 59–70. This was the case throughout the wider Atlantic world. See, for example, Catherine Armstrong, *Writing North America in the Seventeenth Century: English Representations in Print and Manuscript* (Aldershot, UK: Ashgate, 2007), 30.

123. Ogilvie, *The Science of Describing*, 214. See also Lorraine Daston and Peter Galison, *Objectivity* (New York: Zone Books, 2007), ch. 1.

124. Nicolas, *Codex Canadensis*, 284.

125. Ibid.

126. See Daston, "The Nature of Nature in Early Modern Europe."

127. *JR*, 16:191–93.

128. John Steckley, *Words of the Huron* (Waterloo: Wilfrid Laurier University Press, 2007), 113–49.

129. *JR*, 38:243.

130. Ibid., 56:123.

131. These anxieties have been well-documented in other American colonial contexts. See Rebecca Earle, *The Body of the Conquistador: Food, Race, and the Colonial Experience in Spanish America, 1492–1700* (Cambridge: Cambridge University Press, 2012), 28–31, 148–53; Trudy Eden, *The Early American Table: Food and Society in the New World* (DeKalb: Northern Illinois University Press, 2008), ch. 4; LaCombe, *Political Gastronomy*, 56–58.

132. Sagard, *Le Grand Voyage*, 196.

133. *JR*, 6:271–73.

134. Ibid., 18:17.

135. Ibid., 15:167.

136. Nicolas de Ville, *Histoire des plantes de l'Europe, et des plus usitées qui viennent d'Asie, d'Afrique & d'Amérique* (Lyon: Nicolas de Ville, 1689), 32.

137. Louis Liger, *Moyens faciles pour rétablir en peu de temps l'abondance de toutes sortes de grain de fruits dans le royaume* (Paris: Chez Charles Huguier, 1709), 4.

138. Ibid.

139. Pouliot, "Premières pages du journal des Jésuites de Québec," 116.

140. The quoted passage is from Book 1, lines 145–46. Virgil, *Georgics* (Oxford: Oxford University Press, 2006), 10; "labor conquers": David Scott Wilson-Okamura, *Virgil in the Renaissance* (Cambridge: Cambridge University Press, 2010), 77; "toil and the pinch": R. Jenkyns, "Labor Improbus," *Classical Quarterly* 43, no. 1 (1993): 243–48.

141. Aude Doody, "Virgil the Farmer? Critiques of the *Georgics* in Columella and Pliny," *Classical Philology* 102, no. 2 (2007): 180.

142. Mauro Ambrosoli, *The Wild and the Sown: Botany and Agriculture in Western Europe, 1350–1850* (Cambridge: Cambridge University Press, 1997), 67–68.

143. Philip Thibodeau, *Playing the Farmer: Representations of Rural Life in Vergil's Georgics* (Berkeley: University of California Press, 2011), 1–2. Timothy Sweet argues that the georgic was crucial for representing early American environments within the Anglo-Atlantic world. See Timothy Sweet, *American Georgics: Economy and Environment in Early American Literature* (Philadelphia: University of Pennsylvania Press, 2002).

144. Pioffet, *La tentation*, 129.

145. This was unique neither to Virgil nor to his *Georgics*. See, for numerous examples of the complicated relationship between classical texts and Christianity, Anthony Grafton, *Defenders of the Text: The Traditions of Scholarship in an Age of Science, 1450–1800* (Cambridge, Mass.: Harvard University Press, 1991).

146. Thibodeau, *Playing the Farmer*, 7.

CHAPTER 3

1. Innu is "the collective name for the Montagnais-Naskapi." Olive Patricia Dickason, *Canada's First Nations: A History of Founding Peoples from Earliest Times* (Norman: University of Oklahoma Press, 1992), 102.

2. *JR*, 8:57–59.

3. See, for an example of this treatment, Blackburn, *Harvest of Souls*; Maureen F. O'Meara, "Planting the Lord's Garden in New France: Gabriel Sagard's 'Le Grand Voyage' and 'Histoire du Canada,'" *Rocky Mountain Review of Language and Literature* 46, no. 1/2 (1992): 11–24.

4. *JR*, 3:11.

5. Ibid.

6. Ibid.

7. Ibid., 19:37.

8. Ibid., 19:27.

9. Ibid., 20:69–71.

10. This language is inspired by Sayre, "Cultivating Soils and Souls."

11. This is a point made by Cornelius J. Jaenen, "Problems of Assimilation in New France, 1603–1645," *French Historical Studies* 4, no. 3 (1966): 266. Peter Goddard writes that "the conversion of native peoples in New France . . . reflected the missionary project in an age of turmoil."

Peter A. Goddard, "Converting the 'Sauvage': Jesuit and Montagnais in Seventeenth-Century New France," *Catholic Historical Review* 84, no. 2 (1998): 222.

12. See Belmessous, "Assimilation"; Havard, "Les forcer."

13. Dorsey, for example, argues that the Jesuit demand to be "all things to all men in order to win all to Jesus Christ" encouraged considerable flexibility in determining the relationship between external and internal conversion. Peter A. Dorsey, "Going to School with Savages: Authorship and Authority Among the Jesuits of New France," *William and Mary Quarterly* 55, no. 3 (1998): 399.

14. Dominique Deslandres writes, "America in the early times of French colonization constituted a veritable laboratory where an entire array of conversion methods was tested." Deslandres, *Croire et faire croire*, 208. Goddard has likewise characterized New France as a laboratory for Jesuits. Goddard, "Converting the 'Sauvage,'" 219. Sarah Rivett has recently also shown that this was an intense period of investigation of and experimentation with conversion in the Anglophone Atlantic. Sarah Rivett, *The Science of the Soul in Colonial New England* (Chapel Hill: University of North Carolina Press, 2011).

15. Karen Kupperman suggests that something similar occurred in Anglo-American colonies. Kupperman, *Indians and English*, 3.

16. James P. Ronda, "The European Indian: Jesuit Civilization Planning in New France," *Church History* 41, no. 3 (1972): 386.

17. George Stanley, "The Policy of 'Francisation' as Applied to the Indians During the Ancien Regime," *Revue d'histoire de l'Amérique française* 3, no. 3 (1949): 178.

18. Caroline Galland, *Pour la gloire de Dieu et du roi: Les récollets en Nouvelle-France aux XVIIe et XVIIIe siècles* (Paris: Cerf, 2012), 284–85.

19. On the metaphorical significance of agricultural language, see Blackburn, *Harvest of Souls*, ch. 3.

20. *JR*, 10:111.

21. Ibid., 10:103.

22. Ibid., 10:149.

23. See, for example, references to the following. Harvest: ibid., 8:217, 15:147, 19:77; feasts: 7:97, 133, 9:109, 13:29, 43; vineyards: 7:227, 243, 17:125, 19:179.

24. Caroline Galland writes that this vision was already articulated at the very beginning of the Récollet mission. See Galland, *Pour la gloire de Dieu et du roi*, 286. See also Emma Anderson, *The Betrayal of Faith: The Tragic Journey of a Colonial Native Convert* (Cambridge, Mass.: Harvard University Press, 2007), 72–73; Carile, *Le regard entravé*, 16.

25. Chrestien Le Clercq, *Premier établissement de la foy dans la Nouvelle France* (Paris: Amable Auroy, 1691), 285–86.

26. Ibid., 223; Alain Beaulieu, *Convertir les fils de Caïn: Jésuites et amérindiens nomades en Nouvelle-France, 1632–1642* (Québec: Nuit Blanche, 1990), 16.

27. Anderson, *The Betrayal of Faith*, 78.

28. *JR*, 2:139.

29. Ronda, "The European Indian," 385–86. In Acadia, the debate over the readiness of Mi'kmaq peoples to accept Catholicism drove a rupture between Jesuit missionaries and the priest Jessé Fléché. Matteo Binasco, "Few, Uncooperative, and Endangered: The Troubled Activity of the Roman Catholic Missionaries in Acadia (1610–1710)," *Reformation & Renaissance Review: Journal of the Society for Reformation Studies* 8, no. 3 (2006): 326–27.

30. Amy Wyngaard explains that "there is no question that, from the outset, the representation of rural life in seventeenth-century France was fraught with social and ideological

tensions." Amy S. Wyngaard, *From Savage to Citizen: The Invention of the Peasant in the French Enlightenment* (Newark: University of Delaware Press, 2004), 17. See also Liana Vardi, "Imagining the Harvest in Early Modern Europe," *American Historical Review* 101, no. 5 (1996): 1357–97.

31. Marc Jetten, *Enclaves amérindiennes: Les "réductions" du Canada, 1637–1701* (Sillery: Septentrion, 1994), 40.

32. *JR*, 1:85.

33. Ibid., 2:165.

34. This complementarity was explicitly noted; see, for example, ibid., 21:239. Gabriel Sagard also suggests that early colonists at Québec adapted to the trade between Algonquian and Iroquoian communities. See Sagard, *Le Grand Voyage*, 145–46. For academic discussion of this complementarity, see Sioui, *Huron-Wendat*, 65; Bruce G. Trigger, *The Children of Aataentsic: A History of the Huron People to 1660* (Kingston: McGill-Queen's University Press, 1987), 62–63.

35. A great deal of recent research has demonstrated that, while mobile, the Algonquians on the Saint Lawrence migrated seasonally between ecological niches. See Daniel Castonguay, "Les impératifs de la subsistance chez les Montagnais de la Traite de Tadoussac (1720–1750)," *Recherches amérindiennes au Québec* 19, no. 1 (1989): 17–30; idem, "L'exploitation du loup-marin et son incidence sur l'occupation de la côte par les Montagnais de la traite de Tadoussac, au XVIIIe siècle," *Recherches amérindiennes au Québec* 33, no. 1 (2003): 61–72; Harvey Feit, "Les territoires de chasse algonquiens avant leur 'découverte'? Études et histoires sur les tenures, les incendies de forêt et la sociabilité de la chasse," *Recherches amérindiennes au Québec* 34, no. 3 (2004): 5–21.

36. James Axtell, *The Invasion Within: The Contest of Cultures in Colonial North America* (Oxford: Oxford University Press, 1985), 62. Studies of Jesuit strategy in New France have emphasized this adaptation. See, for example, Deslandres, *Croire et faire croire*, 307.

37. *JR*, 6:147.

38. On the poverty of these peoples, see ibid., vol. 6, passim. Le Jeune regularly described Native Americans as poor, where they were either nomadic or sick. Ibid., 8:55 ("idlers"). Peter Goddard explains that evangelization was aimed at equipping Natives to establish "control over what was perceived as depraved human nature." Goddard, "Converting the 'Sauvage,'" 220.

39. Allan Greer suggests that it was learning about aboriginal cultures that pushed some Jesuits to the American mission. Jesuits such as Chauchetière learned a great deal about Canada and understood the opportunity for martyrdom at the hands of aboriginal peoples before they left France. Allan Greer, *Mohawk Saint: Catherine Tekakwitha and the Jesuits* (Oxford: Oxford University Press, 2005), 80–82.

40. See Victor Egon Hanzeli, *Missionary Linguistics in New France: A Study of Seventeenth- and Eighteenth-Century Descriptions of American Indian Languages* (The Hague: Mouton, 1969). Margaret Leahey cites the examples of the Jesuits Noüe and Chabanel to counteract the image that every Jesuit who entered an American mission soon mastered the local language in "'Comment peut un muet prescher l'evangile?' Jesuit Missionaries and the Native Languages of New France," *French Historical Studies* 19, no. 1 (1995): 105–31.

41. John E. Bishop, "Comment dit-on tchistchimanisi8 en français? The Translation of Montagnais Ecological Knowledge in Antoine Silvy's *Dictionnaire montagnais-français* (ca. 1678–1684)" (master's thesis, Memorial University of Newfoundland, 2006), 32–33. True also argues that Jesuits foregrounded their knowledge of indigenous languages. See True, *Masters and Students*, ch. 3.

42. Leahey, "Comment peut un muet prescher l'evangile?" 124; Hanzeli, *Missionary Linguistics in New France*, 49.

43. Londa Schiebinger's *Plants and Empire* is the most vocal advocate of deploying Mary Louise Pratt's notion of a "contact zone" in the study of early modern Atlantic science. While the definition of a biocontact zone involves more than language, Schiebinger has highlighted linguistic mediation as a limiting factor in the exchange of knowledge between Natives, enslaved Africans, and European and colonial naturalists. Schiebinger writes that "noise—or intellectual interference—in biocontact zones was often deafening. Loudest perhaps was the cacophony of languages. Europeans only scratched the surface of local peoples' knowledges of plants and remedies because they were often unable or unwilling to speak local languages." Schiebinger, *Plants and Empire*, 83–84. For an overview of aboriginal medicine in North America, see Virgil J. Vogel, *American Indian Medicine* (Norman: University of Oklahoma Press, 1970).

44. On the tensions caused by Jesuit celibacy, see Richard White, *The Middle Ground: Indians, Empires, and Republics in the Great Lakes Region, 1650–1815* (Cambridge: Cambridge University Press, 1991), 60–61; Trigger, *The Children of Aataentsic*, 360–61. Jesuits adopted aboriginal kinship metaphors when preaching to potential converts and in trying to situate themselves into the social fabric of indigenous communities. See John Steckley, "The Warrior and the Lineage: Jesuit Use of Iroquoian Images to Communicate Christianity," *Ethnohistory* 39, no. 4 (1992): 494–502.

45. Sagard, *Le Grand Voyage*, 271.

46. The ability to learn from indigenous peoples was heavily politicized in the contexts of disputes between Récollets and Jesuits over the Society of Jesus' monopoly on the mission. Jesuits laid claim to the privilege, in part, based upon their efforts (and the challenges) of learning indigenous languages. See True, *Masters and Students*, ch. 2.

47. Sagard, *Dictionaire de la langve Hvronne*, n.p.

48. Ibid. This word most likely actually referred specifically to corn. Sioui, *Huron-Wendat*, 181.

49. Pierre-Joseph-Marie Chaumonot, *Dictionnaire huron*, [before 1697] sme 13/ms-062, 5, La Collection de manuscrits, Séminaire de Québec, Centre de référence de l'Amérique française, Québec; idem, *French Huron Dictionary and Vocabulary*, Codex Ind 12, 12, John Carter Brown Library, Providence, R.I.

50. Jacques Bruyas, *Radices verborum iroquaeorum* (New York: Cramoisy Press, 1862), 32.

51. Sébastien Rasles, *A Dictionary of the Abnaki Language in North America* (Cambridge: Charles Folsom, 1833), 386.

52. Chrestien Leclercq, *Nouvelle relation de la Gaspésie* (Montréal: Presses de l'Université de Montréal, 1999), 283.

53. Sagard, *Le Grand Voyage*, 159.

54. The intended audience of Jesuit writings was never simple. See Florence C. Hsia, *Sojourners in a Strange Land: Jesuits and Their Scientific Missions in Late Imperial China* (Chicago: University of Chicago Press, 2011), 16–18. Luke Clossey similarly points out that the first Jesuit letters from foreign missions were sent in multiple copies and were distributed to multiple audiences. Luke Clossey, *Salvation and Globalization in the Early Jesuit Missions* (Cambridge: Cambridge University Press, 2011), 195. For examples of recent work that has analyzed the impact of audience on the writing of the Jesuit *Relations*, see McShea, "Presenting the 'Poor Miserable Savage' to French Urban Elites"; True, "Maistre et Escolier," 60; Worcester, "A Defensive Discourse," 2–3.

55. Sagard, *Le Grand Voyage*, 152–54. For examples in the first decade of the Jesuit *Relations*, see *JR*, 7:35–37 (Innu housing), 7:97 (tools), 8:103 (Wendat housing), 8:107 (Wendat food), 8:251 (tools), 10:91 (Innu food), 10:147 (Wendat housing).

56. On "frustrated knowledges," see Neil Safier, "Fruitless Botany: Joseph de Jussieu's South American Odyssey," in *Science and Empire in the Atlantic World*, ed. James Delbourgo and Nicholas Dew (New York: Routledge, 2008), 205.

57. *JR*, 7:33.

58. Allan Greer, "Towards a Comparative Study of Jesuit Missions and Indigenous People in Seventeenth-Century Canada and Paraguay," in *Native Christians: Modes and Effects of Christianity Among Indigenous Peoples of the Americas*, ed. Aparecida Villaça and Robin M. Wright (Burlington, Vt.: Ashgate, 2009), 21–32.

59. *JR*, 12:219–21.

60. Ibid., 12:221.

61. Deslandres, *Croire et faire croire*, 95. On the global worldview of Jesuit missionaries, see also Clossey, *Salvation and Globalization in the Early Jesuit Missions*.

62. Karen L. Anderson, *Chain Her by One Foot: The Subjugation of Women in Seventeenth-Century New France* (London: Routledge, 1991), ch. 6.

63. Jean-François Lozier, "In Each Other's Arms: France and the St. Lawrence Mission Villages in War and Peace, 1630–1730" (PhD diss., University of Toronto, 2012), 14.

64. *JR*, 20:273.

65. Beaulieu, *Convertir les fils de Caïn*, 20.

66. *JR*, 21:129.

67. Neophytes famously asked the Jesuits if a woman who ran away from these new strictures ought to be "chained by one foot, and whether four days and nights of fasting would be sufficient penance for her fault." Ibid., 18:3. See also Anderson, *Chain Her by One Foot*.

68. Jetten, *Enclaves amérindiennes*, 38.

69. *JR*, 6:145.

70. Jetten, *Enclaves amérindiennes*, 42.

71. Muriel Clair has recently called them a "failure." Muriel Clair, "'Seeing These Good Souls Adore God in the Midst of the Woods': The Christianization of Algonquian Nomads in the Jesuit Relations of the 1640s," *Journal of Jesuit Studies* 1, no. 2 (2014): 283.

72. Jetten, *Enclaves amérindiennes*, 9.

73. *JR*, 9:225.

74. Ibid., 9:165–67.

75. Ibid., 18:79.

76. Benjamin Breen has shown that this discourse was later resisted by indigenous peoples. Benjamin Breen, "'The Elks Are Our Horses': Animals and Domestication in the New France Borderlands," *Journal of Early American History* 3, no. 2–3 (2013): 181–206. See also Christopher M. Parsons, "Apprendre en apprivoisant: La domestication comme lieu de rencontre dans la France coloniale d'Amérique du Nord," in *Penser l'Amérique: De l'observation à l'inscription*, ed. Nathalie Vuillemin and Thomas Wien (Oxford: Voltaire Foundation, 2017), 143–64.

77. Clair, "'Seeing These Good Souls Adore God in the Midst of the Woods,'" 291.

78. Kenneth M. Morrison, *The Solidarity of Kin: Ethnohistory, Religious Studies, and the Algonkian-French Religious Encounter* (Albany: State University of New York Press, 2002), 104.

79. On Wendat agriculture, see Conrad Heidenreich, *Huronia: A History and Geography of the Huron Indians, 1600–1650* (Toronto: McClelland and Stewart, 1971), ch. 6.

80. *JR*, 8:105.

81. For representative descriptions, see Sagard, *Le Grand Voyage*, 179–80, 93–94.

82. Ibid., 171.

83. Marcel Moussette, "Le Canada," in *Archéologie de l'Amérique française*, ed. Marcel Moussette and Gregory A. Waselkov (Montréal: Lévesque éditeur, 2013), 206–8; Gilles Thérien, "L'inscription dans le paysage: Un examen des modes d'habitation en Nouvelle-France depuis le XVIe siècle," *Études françaises* 22, no. 2 (1986): 57–59.

84. Accounts of the gendered division of labor can be found in any survey of Iroquoian and agricultural Algonquian peoples. See Cronon, *Changes in the Land*, 44; Bruce G. Trigger, *The Huron: Farmers of the North* (Toronto: Holt, Rinehart and Winston, 1969), 34. See also Tremblay, *Les Iroquoiens du Saint-Laurent*, 74–77. For a more focused study of the role of Haudenosaunee women in both the production and distribution of agricultural products, see Judith K. Brown, "Economic Organization and the Position of Women Among the Iroquois," *Ethnohistory* 17, no. 3/4 (1970): 151–67.

85. Dean Snow, who wrote that Iroquoian agriculture was maintained by "a society of women in each village," which "maintained the ceremonies needed to propitiate the spirits of the three sisters," demonstrates that Jesuit observations were well-founded. See Dean R. Snow, *The Iroquois* (Oxford: Blackwell, 1994), 69. French observers generally misunderstood this division, however, and many saw aboriginal women as the slaves of indolent aboriginal men. Havard and Vidal, *Histoire de l'Amérique française*, 636.

86. Bruce M. White, "The Woman Who Married a Beaver: Trade Patterns and Gender Roles in the Ojibwa Fur Trade," *Ethnohistory* 46, no. 1 (1999): 123–27. Thomas Vennum suggests, even more specifically, that among the Menominee wild rice was the property of specific totemic clans, the Bear and Sturgeon. See Thomas Vennum, *Wild Rice and the Ojibway People* (St. Paul: Minnesota Historical Society Press, 1988), 59–60.

87. Quoted in Greer, *Mohawk Saint*, 34.

88. Joseph-François Lafitau, *Customs of the American Indians Compared with the Customs of Primitive Times*, trans. William N. Fenton and Elizabeth L. Moore (Toronto: Champlain Society, 1974–77), 2:54–55.

89. Ibid., 2:47.

90. Lewis H. Morgan, *League of the Ho-dé-no-sau-nee or Iroquois* (New York: Dodd, Mead and Company, 1904), 153.

91. For work analyzing the confrontation between Jesuits and aboriginal women, see Anderson, *Chain Her by One Foot*; Carol Devens, *Countering Colonization: Native American Women and Great Lakes Missions, 1630–1900* (Berkeley: University of California Press, 1992). For a more recent moderation of the claim that Jesuits undermined the position of women in aboriginal societies, see Susan Sleeper-Smith, "Women, Kin, and Catholicism: New Perspectives on the Fur Trade," *Ethnohistory* 47, no. 2 (2000): 423–52; idem, *Indian Women and French Men: Rethinking Cultural Encounter in the Western Great Lakes* (Amherst: University of Massachusetts Press, 2001).

92. *JR*, 54:143.

93. Steckley, *Words of the Huron*, 113.

94. Barbara Mann, "The Lynx in Time: Haudenosaunee Women's Traditions and History," *American Indian Quarterly* 21, no. 3 (1997): 425.

95. Ibid.

96. Claude Chauchetière, *Narration de la mission du Sault depuis sa fondation jusqu'en 1686* (Bordeaux: Archives départmentales de la Gironde, 1984), n.p.

97. There are numerous examples of Jesuits who joined seminomadic aboriginal peoples as they broke into smaller family groups and hunted through the winter; see *JR*, 60:223–25, for

example. Jesuit resistance to joining similar trips in Iroquoian societies limited access to the ecological knowledge of these communities; see ibid., 54:117, for example.

98. Stephen G. Monckton, *Huron Paleoethnobotany* (Toronto: Ontario Heritage Foundation, 1992), 107–25.

99. The most thorough discussion of plant use among contemporary and historical aboriginal communities of the Great Lakes is Richard Yarnell, *Aboriginal Relationships Between Culture and Plant Life in the Upper Great Lakes Region* (Ann Arbor: University of Michigan Press, 1964). For Great Lakes Algonquian communities, see Frances Densmore, *Strength of the Earth: The Classic Guide to Ojibwe Uses of Native Plants* (St. Paul: Minnesota Historical Society Press, 2005); James W. Herrick, *Iroquois Medical Botany* (Syracuse, N.Y.: Syracuse University Press, 1995); F. W. Waugh, *Iroquois Foods and Food Preparation* (Ottawa: Canada Dept. of Mines, Geological Survey, 1916).

100. Stephen Monckton argues, for example, that the historical geographer Conrad Heidenreich's reconstruction of the Wendat diet vastly overemphasized the contribution of corn to aboriginal diets because of his reliance on Jesuit sources that made the same mistake. Monckton's estimate of wild fruit consumption among the Wendat is three times larger than Heidenreich's. Monckton, *Huron Paleoethnobotany*, 112–17. See also Heidenreich, *Huronia*, ch. 6.

101. Lafitau, *Customs of the American Indians*, 1:56–63.

102. Pierre-François-Xavier de Charlevoix, *Journal d'un voyage fait par ordre du Roi dans l'Amérique septentrionale* (Montréal: Presses de l'Université de Montréal, 1994), 2:942.

103. Ibid., 2:943.

104. *JR*, 39:247.

105. Ibid., 59:129. On the environmental history of indigenous peoples in Illinois during this period, see Robert Michael Morrissey, "Bison Algonquians: Cycles of Violence and Exploitation in the Mississippi Valley Borderlands," *Early American Studies* 13, no. 2 (2015): 309–40; Morgan, *Land of Big Rivers*.

106. *JR*, 59:69.

107. Ibid., 59:71.

108. Paul Nadasdy, *Hunters and Bureaucrats: Power, Knowledge, and Aboriginal-State Relations in the Southwest Yukon* (Vancouver: University of British Columbia Press, 2003), 60. Quoted in Bishop, "Comment dit-on tchistchimanisi8 en français?" 20.

109. *JR*, 64:137.

110. Ibid., 38:243. For more on the failure to acknowledge aboriginal use of "wild" plants, see Chapter 1. For a more in-depth study of cognitive gaps and their effect on perceptions and use of colonial environments, see Tsing, *Friction*, ch. 5.

111. The question of whether agricultural crops were native to eastern North America or were entirely imported from the Mexico region has been debated for some time. Yarnell suggests that plant domestication began in the late archaic in the east and included plants such as "giant ragweed, two wild beans, Jerusalem artichoke, groundnut, maypop, black nightshade, purslane, carpetweed, a spurge, and perhaps other species." Richard Yarnell, "The Importance of Native Crops During the Late Archaic and Woodland Periods," in *Foraging and Farming in the Eastern Woodlands*, ed. C. Margaret Scarry (Gainesville: University Press of Florida, 1993), 22. More recently, genetic evidence has been used to demonstrate conclusively that marshelder, chenopod, squash, and sunflower were indigenous to eastern North America. Bruce D. Smith, "Eastern North America as an Independent Center of Plant Domestication," *Proceedings of the National Academy of Science* 103, no. 33 (2006): 12223–28. Archaeologists have suggested that

fires were locally important and carefully set in areas with readily available water and away from settlements. See James S. Clark and P. Daniel Royall, "Local and Regional Sediment Charcoal Evidence for Fire Regimes in Presettlement North-eastern North America," *Journal of Ecology* 84, no. 3 (1996): 365–82.

112. *JR*, 55:167; Nicolas Perrot, *Moeurs, coutumes et religion des sauvages de l'Amérique septentrionale* (Montréal: Presses de l'Université de Montréal, 2004), 255.

113. Gayle J. Fritz, "Levels of Native Biodiversity in Eastern North America," in *Biodiversity and Native America*, ed. Paul E. Minnis and Wayne J. Elisens (Norman: University of Oklahoma Press, 2000), 224; John L. Riley, *The Once and Future Great Lakes Country: An Ecological History* (Kingston: McGill-Queen's University Press, 2013), 156–59.

114. A. Irving Hallowell, "Ojibwa Ontology, Behavior, and Worldview," in *Contributions to Anthropology: Selected Papers of A. Irving Hallowell*, ed. Raymond D. Fogelson (Chicago: University of Chicago Press, 1976), 368.

115. Ibid., 369.

116. Bruce M. White, "Encounters with Spirits: Ojibwa and Dakota Theories About the French and Their Merchandise," *Ethnohistory* 41, no. 3 (1994): 380–81.

117. *JR*, 5:165. Kenneth Morrison treats this and other encounters of this sort as, in part, a collision of French and indigenous conceptions of what made up the animate world. Morrison, *The Solidarity of Kin*, 113–14.

118. Morrison, *The Solidarity of Kin*, 121.

119. See, for representative examples, *JR*, 9:211, 67:157–59.

120. Ibid., 65:63.

121. Several scholars have discussed colonial faith in the curative powers of local saints, for example. See Greer, *Mohawk Saint*, ch. 7; Timothy G. Pearson, *Becoming Holy in Early Canada* (Kingston: McGill-Queen's University Press, 2014), ch. 6.

122. *JR*, 23:57.

123. Ibid., 20:81.

124. Examples of these ceremonies abound. See Herrick, *Iroquois Medical Botany*, 2; Steckley, *Words of the Huron*, 113–15; Elisabeth Tooker, *An Ethnography of the Huron Indians, 1615–1649* (Syracuse, N.Y.: Syracuse University Press, 1991), 62.

125. Tobacco was understood to provide a bridge between the human and non-human world. See Parsons, "Of Natives, Newcomers, and Nicotiana."

126. This is reminiscent of what anthropologist Richard Nelson has referred to as "the watchful world" of the Koyukon peoples. Richard K. Nelson, *Make Prayers to the Raven: A Koyukon View of the Northern Forest* (Chicago: University of Chicago Press, 1983), 14–33.

127. Lafitau, *Customs of the American Indians*, 1:218.

128. Tooker, *An Ethnography of the Huron Indians*, 90–91, 109. See also Kenn Pitawanakwat and Jordan Paper, "Communicating the Intangible: An Anishnaabeg Story," *American Indian Quarterly* 20, no. 3–4 (1996): 451–65.

129. William M. Beauchamp, *Iroquois Folk Lore: Gathered from the Six Nations of New York* (Port Washington, N.Y.: Ira J. Friedman, Inc., 1922), 49.

130. Ibid., 50.

131. Ibid., 50–51.

132. Marius Barbeau, *Mythologie huronne et wyandotte: Avec en annexe les textes publieés antérieurement*, trans. Stephan Dupont (Montréal: Presses de l'Université de Montréal, 1994), 94.

133. Herrick, *Iroquois Medical Botany*, 35–36.

134. Wendy Makoons Geniusz, *Our Knowledge Is Not Primitive: Decolonizing Botanical Anishinaabe Teachings* (Syracuse, N.Y.: Syracuse University Press, 2009), 64.

135. On the importance of secrecy, see William Wykoff, "Botanique et Iroquois dans la vallée du St-Laurent," *Anthropologie et Sociétés* 2, no. 3 (1978): 157–62. See also William N. Fenton, *Contacts Between Iroquois Herbalism and Colonial Medicine* (Seattle: Shorey Book Stores, 1971), 523.

136. See William Fenton's ethnobotanical notebooks for the years 1938, 1939, and 1940, catalogued respectively as "Plants," "Plants. Field Notes 1," and "Plants. Field Notes 2" in Box 6-Series 5, Ms. Coll. 20, William Fenton Papers, American Philosophical Society, Philadelphia.

137. Ibid.; Fenton, *Contacts Between Iroquois Herbalism and Colonial Medicine*, 507.

138. William Fenton, "Iroquois Use of Plants as Medicines," Ms. [1954], Box 8, Ms. Coll. 20, William Fenton Papers, American Philosophical Society, Philadelphia.

139. Ibid.

140. Geniusz, *Our Knowledge Is Not Primitive*, 66.

141. *JR*, 30:23.

142. Nicolas, *Codex Canadensis*, 298.

143. *JR*, 59:101.

144. Charlevoix, *Journal d'un voyage*, 1:498.

145. Ibid., 2:622.

146. Robert Morrissey suggests that this was not limited to the exchange of botanical knowledge but was part of a broader pattern that saw tension between missionaries and aboriginal communities increase as Jesuits gained fluency in indigenous languages. See Robert Michael Morrissey, "'I Speak It Well': Language, Cultural Understanding, and the End of a Missionary Middle Ground in Illinois Country, 1673," *Early American Studies* 9, no. 3 (2011): 617–48.

147. References to conflicts between missionaries and shamans are numerous. See Axtell, *The Invasion Within*, 77–88; Blackburn, *Harvest of Souls*, ch. 5; Trigger, *The Children of Aataentsic*, 391, 407, 530, 707.

148. *JR*, 13:105.

149. Ibid.

150. The study of the cultural production of ignorance has been introduced to the study of early modern Atlantic science as "agnotology" in the work of Londa Schiebinger. See Schiebinger, *Plants and Empire*, 226–41.

151. *JR*, 14:125. See also Allan Greer, "The Exchange of Medical Knowledge Between Natives and Jesuits in New France," in *El saber de los jesuitas, historias naturales y el Nuevo Mundo*, ed. Domingo Ledezma and Luis Millones Figueroa (Frankfurt: Vervuert-Iberoamericana, 2005), 135–46. This was not unique to the missions to New France. See Fernando Cervantes, *The Devil in the New World: The Impact of Diabolism in New Spain* (New Haven, Conn.: Yale University Press, 1994), ch. 1.

152. Chris Parsons, "Medical Encounters and Exchange in Early Canadian Missions," *Scientia Canadensis* 31, no. 1–2 (2008): 57.

153. Peter Goddard, "The Devil in New France: Jesuit Demonology, 1611–50," *Canadian Historical Review* 78, no. 1 (1997): 41. See also J. Michelle Molina, *To Overcome Oneself: The Jesuit Ethic and Spirit of Global Expansion, 1520–1767* (Berkeley: University of California Press, 2013), 49–50.

154. Lafitau, *Customs of the American Indians*, 1:227.

CHAPTER 4

1. Gédéon de Catalogne, "Report on the Seigniories and Settlements in the Districts of Quebec, Three Rivers, and Montreal, by Gedeon de Catalogne/Engineer, November 7, 1712," in *Documents Relating to the Seigniorial Tenure in Canada, 1598–1854*, ed. William Bennett Munro (Toronto: Champlain Society, 1908), 95. One historian has, since then, "doubted whether any one [*sic*] save Catalogne had at any stage of the French régime a personal knowledge of conditions in every seigniory of the colony." William Bennett Munro, "Historical Introduction," in *Documents Relating to the Seigniorial Tenure in Canada, 1598–1854*, ed. William Bennett Munro (Toronto: Champlain Society, 1908), lix.

2. Jean-François Palomino, "Pratiques cartographiques en Nouvelle-France: La prise en charge de l'état dans la description de son espace colonial à l'orée du xviiie siècle," *Lumen* 31 (2012): 22. See also the excellent edited collection by Raymonde Litalien, Denis Vaugeois, and Jean-François Palomino, *La mesure d'un continent: Atlas historique de l'Amérique du nord, 1492–1814* (Sillery: Septentrion, 2007).

3. See, for more information, "Catalonge, Gédéon (de)," *Dictionary of Canadian Biography*, http://www.biographi.ca/en/bio/catalogne_gedeon_2F.html. The intendant was a position introduced in New France after Louis XIV took direct control of the colony and was inspired by similar administrative reform in France. As Allan Greer writes, "Intendants were normally well-qualified administrators with legal training, financial expertise, and prior experience in the French naval bureaucracy. An intendant was particularly charged with overseeing New France's judicial apparatus and handling the government's accounts; otherwise, he was supposed to cooperate with the governor in administering the colony." Allan Greer, *The People of New France* (Toronto: University of Toronto Press, 1997), 44.

4. Jean-François Palomino, "Portrait d'un cartographe: Jean-Baptiste Franquelin," in *La mesure d'un continent: Atlas historique de l'Amérique du nord, 1492–1814*, ed. Raymonde Litalien, Denis Vaugeois, and Jean-François Palomino (Sillery: Septentrion, 2007), 104.

5. Space has been a productive analytical category for historians of empire throughout the Atlantic world and has been particularly important for demonstrating the contested and contingent nature of some of the most basic categories that organized colonial experience. The history of the efforts to come to terms with New France during this period suggests that a "spatial history of empire" was also an environmental history of empire. Ecology was inextricably bound up in the creation of what historian Elizabeth Mancke has referred to as the "spaces of power" in early America. As Mancke explains, "By 'spaces of power' I mean systems of social power, whether economic, political, cultural or military, that we can describe functionally and spatially." Elizabeth Mancke, "Spaces of Power in the Early Modern Northeast," in *New England and the Maritime Provinces, Connections and Comparisons* (Kingston: McGill-Queen's University Press, 2005), 32. See also Jeffers Lennox, "A Time and a Place: The Geography of British, French, and Aboriginal Interactions in Early Nova Scotia, 1726–44," *William and Mary Quarterly* 72, no. 3 (2015): 423–60; Juliana Barr and Edward Countryman, eds., *Early American Studies: Contested Spaces of Early America* (Philadelphia: University of Pennsylvania Press, 2014). We should equally remember the complexity and continuity of indigenous geographies in this region. See, for studies of the construction and experience of indigenous spaces in the Great Lakes region, Heidi Bohaker, "'Nindoodemag': The Significance of Algonquian Kinship Networks in the Eastern Great Lakes Region, 1600–1701," *William and Mary Quarterly* 63, no. 1 (2006): 23–52; Michael J. Witgen, *An Infinity of Nations: How the Native New World Shaped Early North America* (Philadelphia: University of Pennsylvania Press, 2012).

6. Although it has often been historians and scholars of maps and mapmaking who have studied the history of representations of space, literature scholars have also demonstrated the importance of the literary in representing space. See Martin Brückner and Hsuan L. Hsu, eds., *American Literary Geographies: Spatial Practice and Cultural Production, 1500–1900* (Newark: University of Delaware Press, 2007).

7. "Mémoire de Gédéon de Catalogne sur le Canada," C11A, 33:210, Archives nationales d'outre-mer (hereafter ANOM).

8. Ibid., 213.

9. Ibid.

10. See Chapter 1.

11. Nicolas, *Codex Canadensis*, 280.

12. Lahontan, *Oeuvres complètes* (Montréal: Presses de l'Université de Montréal, 2000), 593–94.

13. "Mémoire de Gédéon de Catalogne sur le Canada," 214.

14. Catalogne, "Report on the Seigniories and Settlements," 117.

15. See Chapter 1. See also Ferland, *Bacchus en Canada*.

16. Brazeau, *Writing a New France*, ch. 1.

17. Nicolas, *Codex Canadensis*, 287.

18. Charlevoix, *Journal d'un voyage*, 1:208.

19. J. Aulneau, "Lettre de Père Aulneau à sa mère (10 October 1734)," *Rapport de l'archiviste de la province de Québec* (1927): 263.

20. *JR*, 64:133.

21. For a discussion of the climatic differences between Québec and Montréal, see Thomas Wien, "'Les travaux pressants': Calendrier agricole, assolement et productivité au Canada au XVIIIe siècle," *Revue d'histoire de l'Amérique française* 43, no. 4 (1990): 542. For a careful study of the gardens and orchards at Montréal, and a discussion of the properties of religious institutions, see Dépatie, "Jardins et vergers à Montréal au XVIIIe siècle," 234–35.

22. This corresponds with a broader interest in the metaphor of grafting in eighteenth-century France. Giulia Pacini, "Grafts at Work in Late Eighteenth-Century French Discourse and Practice," *Eighteenth-Century Life* 34, no. 2 (2010): 1–22.

23. "Les journaux de M. de Léry (26 Mai 1749)," *Rapport de l'archiviste de la province de Québec* (1927): 335. On fruit growing at Detroit, see Guillaume Teasdale, "The French of Orchard Country: Territory, Landscape, and Ethnicity in the Detroit River Region, 1680s–1810s" (PhD diss., York University, 2010), ch. 5.

24. "Relation de la Louisianne ou Mississipi. Ecrite à une dame, par un officier de marine," [c. 1735], Ms. 530, p. 232, Edward E. Ayer Collection, Newberry Library, Chicago.

25. Ibid., 233.

26. Ibid., 236.

27. My thanks to Laura Green for working through this part of the text with me.

28. Ibid.; Dumont de Montigny also notes the success of introduced Muscat grapes. Jean-François-Benjamin Dumont de Montigny, *Regards sur le monde atlantique, 1715–1747* (Sillery: Septentrion, 2008), 401.

29. For a brief overview of linguistic borrowing in New France, see Marthe Faribault, "L'emprunt amérindien en français de la Nouvelle-France: Solutions de quelques problèmes d'etymologie," in *Français du Canada-Français de France: Actes du cinquième colloque international de Bellême du 5 au 7 juin 1997*, ed. Marie-Rose Simoni-Aurembou (Tübingen: Max Niemeyer Verlag, 2000).

30. The rough chronology (and particularly early references to medlars that I would not have connected to persimmon) was made in C. H. Briand, "The Common Persimmon (Diospyros virginiana L.): The History of an Underutilized Fruit Tree (16th–19th Centuries)," *Huntia* 12, no. 1 (2005): 71–89.

31. Marc Lescarbot, *History of New France*, trans. W. L. Grant (Toronto: Champlain Society, 1907), 257. Even in the early eighteenth century, other colonial naturalists continued to affix the plant with the name of already familiar plants. The English naturalist Marc Catesby, for example, called the plant the Guajacana in his *Natural History of Carolina, Florida and the Bahama Islands*, naming the plant after a related species that had been brought to Europe from Africa in the early seventeenth century. Mark Catesby, *The Natural History of Carolina, Florida and the Bahama Islands* (London, 1731), 2:76.

32. Joutel, *Journal historique*, 154.

33. *JR*, 65:115–17.

34. "Mémoire de Gédéon de Catalogne sur le Canada," 213.

35. Ibid., 212.

36. The daily contact between French and indigenous peoples in many parts of French North America—and particularly at the periphery of empire in the *pays d'en haut*—often also facilitated the encounter with indigenous flora as foods, botanical commodities, and wild and cultivated live plants. For recent studies of cultural contact and exchange in these regions that touch upon botanical exchange, see Arnaud Balvay, *L'Épée et la plume: Amérindiens et soldats des troupes de la marine en Louisiane et au Pays d'en Haut (1683–1763)* (Québec: Presses de l'Université Laval, 2006), ch. 8; Havard, *Empire et métissages: Indiens et français dans le Pays d'en Haut* (Québec: Septentrion, 2003), ch. 9. See also Usner, *Indians, Settlers, & Slaves in a Frontier Exchange Economy*.

37. Potier arrived in New France in the 1740s and traveled from Wendat missions in the Saint Lawrence to those in the Great Lakes, at Detroit. See, for more information on his work, Thomas G. M. Peace, "Two Conquests: Aboriginal Experiences of the Fall of New France and Acadia" (PhD diss., York University, 2011), ch. 5.

38. Marthe Faribault, "Le vocabulaire botanique dans les écrits du père Potier (XVIIIe siiécle)," *Humanities Research Group Working Papers (University of Windsor)* 11 (2003): 84. See, for more usage, the entry for "pacane" in the glossary of Etienne Martin de Vaugine de Nuisement. Steve Canac-Marquis and Pierre Rézeau, eds., *Journal de Vaugine de Nuisement (ca. 1765): Un témoignage sur la Louisiane du XVIIIe siècle* (Sainte-Foy: Presses de l'Université Laval, 2005), 116.

39. Faribault, "Le vocabulaire botanique," 84.

40. Robert Vézina, "Le lexique des voyageurs francophones et les contacts interlinguistiques dans le milieu de la traite des pelleteries: Approche sociohistorique, philologique et lexicologique" (PhD diss., Université Laval, 2010), 128.

41. Perrot, *Moeurs*, 268.

42. Pierre Deliette, "Memoir of Pierre Liette on the Illinois Country," in *The Western Country in the 17th Century: The Memoirs of Lamothe Cadillac and Pierre Liette*, ed. Milo Milton Quaife (Chicago: Lakeside Press, 1947), 104.

43. Lucien Campeau, "Les origines du sucre d'érable," *Les cahiers des dix* 45 (1990): 54; Yvon Desloges, *À table en Nouvelle-France: Alimentation populaire, gastronomie et traditions alimentaires dans la vallée laurentienne avant l'avènement des restaurants* (Québec: Septentrion, 2009), 72.

44. Boucher, *Histoire veritable*, 44–45.

45. J. C. B., "Voyage au Canada dans le nord de l'Amerique septentrionale, fait depuis l'an 1751 à 1761," Nouvelles acquisitions françaises, Bibliothèque Nationale de France, Paris, 4156, 165. J. C. B. mentions "le plane, le merisier, le fresne et le noyer" specifically as trees that would produce syrup.

46. Quoted in Réal Ouellet, "Le Sirop d'Erable," in Chrestien Leclercq, *Nouvelle relation de la Gaspésie* (Montréal: Presses de l'Université de Montréal, 1999), 658.

47. Buc'hoz, *Dictionnaire universel des plantes*, 557–61. For a discussion of why American maples transplanted in France did not produce as much syrup, see Henri-Louis Duhamel du Monceau, *Traité des arbres et arbustes qui se cultivent en France en pleine terre* (Paris: H. L. Guerin and L. F. Delatour, 1755), 1:35–36. Linnaeus apparently also tried this in Sweden. See Staffan Müller-Wille, "Walnuts at Hudson Bay, Coral Reefs in Gotland: The Colonialism of Linnaean Botany," in *Colonial Botany: Science, Commerce, and Politics in the Early Modern World*, ed. Londa Schiebinger and Claudia Swan (Philadelphia: University of Pennsylvania Press, 2005), 37–38.

48. Charlevoix, *Journal d'un voyage*, 1:307.

49. J. C. B., "Voyage au Canada," 164.

50. Bernard Boivin, "La flore du Canada en 1708: Étude et publication d'un manuscrit de Michel Sarrasin et Sébastien Vaillant," *Études littéraires* 10, no. 1–2 (1977): 213. Sarrazin's account of four species of Canadian maples was published in "Observations botaniques," *Histoire de l'Académie Royale des Sciences avec les Mémoires de Mathématique & de Physique, pour la même Année* (1730): 65–66 (hereafter *HMARS*).

51. Grenier, *Bréve histoire du régime seigneurial*, 63–68.

52. Conrad Heidenreich and Edward H. Dahl, "Samuel de Champlain's Cartography, 1603–32," in *Champlain: The Birth of French America*, ed. Raymonde Litalien and Denis Vauegeois (Kingston: McGill-Queen's University Press, 2004); Litalien, Vaugeois, and Palomino, *La mesure d'un continent*.

53. See Chapter 1. See also Coates and Degroot, "'Les bois engendrent les frimas et les gelées,'" where the examples of this logic are most visible in the early seventeenth century.

54. Catalogne, "Report on the Seigniories and Settlements," 106.

55. Ibid., 104.

56. Ibid., 116.

57. Courville, *Le Québec*, 125–26. Harris argues that access to good waterways was as least as important as good soil for censitaires. Richard Colebrook Harris, *The Seigneurial System in Early Canada: A Geographical Study* (Kingston: McGill-Queen's University Press, 1984), 127.

58. See, for an overview of the early use of this term, Gervais Carpin, *Histoire d'un mot: L'ethnonyme "canadien" de 1535–1691* (Sillery: Septentrion, 1995).

59. James Pritchard, *In Search of Empire: The French in the Americas, 1670–1730* (Cambridge: Cambridge University Press, 2004), xxi.

60. Carpin, *Histoire d'un mot*; Belmessous, "Etre français en Nouvelle-France"; White, *Wild Frenchmen and Frenchified Indians*.

61. Moussette, "Le Canada," 301–2. Frégault, for example, sees the early eighteenth century as a critical period of self-exploration and cultural formation. Guy Frégault, *La civilisation de la Nouvelle-France, 1713–1744* (Anjou, Québec: Éditions Fides, 2014 [1969]). Frégault and others have focused particular attention on the decades of peace in the early eighteenth century, which have posed particular narrative and archival challenges. See Thomas Wien, "En attendant Frégault: À propos de quelques pages blanches de l'histoire du Canada sous le Régime français," in *De Québec à l'Amérique française: Textes choisis du deuxième colloque de la Commission*

franco-québécoise sur les lieux de mémoire communs, ed. Thomas Wien, Cécile Vidal, and Yves Frenette (Sainte-Foy: Presses de l'Université Laval, 2006).

62. Carpin, *Histoire d'un mot*, 26.

63. Christian Morissonneau, "Dénommer les terres neuves: Cartier et Champlain," *Études littéraires* 10, no. 1–2 (1977): 102.

64. Conrad Heidenreich and K. Janet Ritch, "Champlain and His Times to 1604," in *Samuel de Champlain Before 1604: Des Sauvages and Other Documents Related to the Period*, ed. Conrad Heidenreich and K. Janet Ritch (Montréal: McGill-Queen's University Press, 2010), 92.

65. Carpin, *Histoire d'un mot*, 63. See also Laurier Turgeon, "The Cartier Voyages to Canada (1534–42) and the Beginnings of French Colonialism in North America," in *Charting Change in France Around 1540*, ed. Marian Rothstein (Selinsgrove, Pa.: Susquehanna University Press, 2006).

66. These terms were paired, and "New France" was a term that almost invariably referred to North America. Brazeau, *Writing a New France*, 12.

67. See, for representative examples, the documents assembled in Robert Le Blant and René Baudry, *Nouveaux documents sur Champlain et son Époque* (Ottawa: Public Archives of Canada, 1967).

68. See Champlain, "Carte geographique de la Nouelle Franse," in Champlain, *Works*, vol. 2. Lestringant argues that this was part of Champlain's efforts to assert himself as the founder of New France. Lestringant, "Champlain, Lescarbot et la 'conférence' des histoires."

69. Sagard, *Histoire du Canada*. See also Carpin, *Histoire d'un mot*, 92–93.

70. And it has been an equally productive conflation for some scholars. Greer, *The People of New France*, 3–4.

71. Jean Talon, "Lettre de Talon au ministre Colbert (10 Octobre 1665)," *Rapport de l'archiviste de la province de Québec* (1931): 39.

72. Jean Talon, "Lettre de Talon au ministre Colbert (4 Octobre 1665)," *Rapport de l'archiviste de la province de Québec* (1931): 32.

73. Jaenen calls this a "utopian dream." Cornelius J. Jaenen, "Colonisation compacte et colonisation extensive aux xviie et xviiie siècles en Nouvelle-France," in *Colonies, territoires, sociétés: L'enjeu français*, ed. Alain Saussol and Joseph Zitomersky (Paris: L'Harmattan, 1996), 16.

74. Carpin, *Histoire d'un mot*, 135–36.

75. Boucher, *Histoire veritable*, 4.

76. Palomino, "Portrait d'un cartographe," 106–7.

77. Guillaume Delisle, *Carte du Canada ou de la Nouvelle France et des découvertes qui y ont été faites* (Paris, 1703); Jacques Nicolas Bellin, *Partie orientale de la Nouvelle France ou du Canada* (Paris, 1755). Interestingly, this differs markedly from the 1744 edition of the Bellin map, where New France is used to demarcate roughly the same space. Jacques Nicolas Bellin, *Partie orientale de la Nouvelle France ou du Canada* (Paris, 1744).

78. See Paul W. Mapp, *The Elusive West and the Contest for Empire, 1713–1763* (Chapel Hill: University of North Carolina Press, 2011), chs. 5–8.

79. Jean-Baptiste Colbert, "Lettre du ministre Colbert à Talon (5 Janvier 1666)," *Rapport de l'archiviste de la province de Québec* (1931): 41.

80. Ibid.

81. The literature on the "compact colony" policy is voluminous. See, for an overview of the evolution of the idea, Dale Miquelon, "Jean-Baptiste Colbert's 'Compact Colony Policy'

Revisited: The Tenacity of an Idea," in *Proceedings of the Seventeenth Meeting of the French Colonial Historical Society*, ed. Patricia Galloway (Lanham, Md.: University Press of America, 1993).

82. Quoted in W. J. Eccles, *Canada Under Louis XIV, 1663–1701* (Toronto: McClelland and Stewart, 1964), 105–6.

83. Miquelon, "Jean-Baptiste Colbert's 'Compact Colony Policy' Revisited," 17; idem, "Les Pontchartrain se penchent sur leurs cartes de l'Amérique: Les cartes et l'impérialisme, 1690–1712," *Revue d'histoire de l'Amérique française* 59, no. 1–2 (2005): 53–71.

84. Marcel Trudel, *La seigneurie de la Compagnie des Indes occidentales, 1663–1674* (Montréal: Fides, 1997), 3; Greer, *The People of New France*, 4.

85. Dale Miquelon, *New France, 1701–1744: A Supplement to Europe* (Toronto: McClelland and Stewart, 1987), 10.

86. Allan Greer calls this a "revolution in the life of the St. Lawrence colony." Greer, *Peasant, Lord, and Merchant*, 4. See also Coates, *The Metamorphoses of Landscape and Community in Early Quebec*, 11.

87. Choquette writes, "The *filles du roi* and some of the soldiers proved to be excellent colonists." Leslie Choquette, *Frenchmen into Peasants: Modernity and Tradition in the Peopling of French Canada* (Cambridge, Mass.: Harvard University Press, 1997), 276. See also Jacques Mathieu and Serge Courville, eds., *Peuplement colonisateur aux XVIIe et XVIIIe siècles* (Québec: Cahiers du Célat, 1987); Trudel, *La seigneurie de la Compagnie des Indes occidentales*, 170–73, 265–68.

88. Dechêne, *Habitants and Merchants*, 22–23.

89. Havard and Vidal, *Histoire de l'Amérique française*, 101.

90. Alfred W. Crosby, *Ecological Imperialism: The Biological Expansion of Europe, 900–1900*, new ed. (New York: Cambridge University Press, 2004), 3.

91. Havard, *Empire et métissages*, 52. Jack Warwick dates the use of the term *pays d'en haut* to 1712. Jack Warwick, "Les 'pays d'en haut' dans l'imagination canadienne-française," *Études françaises* 2, no. 3 (1966): 265–93.

92. Havard, *Empire et métissages*, 52.

93. Arnaud Balvay, for example, has analyzed how historical authors translated the transition between the interior and exterior spaces of French forts in the Great Lakes into a transition between "barbarous" and "civilized" spaces. Balvay, *L'Épée et la plume*, ch. 4. Gilles Havard has similarly described lasting anxieties about the "degeneration" of the Frenchmen who traveled west that was already being refuted in the eighteenth century by figures such as Lahontan. Havard, *Empire et métissages*, 535–37.

94. "Royal Instructions given to the Sieur Gaudais, Special Commissioner to Investigate Conditions in New France, May 7, 1663," in *Documents Relating to the Seigniorial Tenure in Canada, 1598–1854*, ed. William Bennett Munro (Toronto: Champlain Society, 1908), 15.

95. Thomas Wien, "Vie et transfiguration du coureur de bois," in *Mémoires de Nouvelle-France: De France en Nouvelle-France*, ed. Philippe Joutard and Thomas Wien (Rennes: Presses Universitaires de Rennes, 2005).

96. Quoted in Miquelon, "Jean-Baptiste Colbert's 'Compact Colony Policy,' Revisited," 14–15.

97. Quoted in Belmessous, "Etre français en Nouvelle-France," 526. On these marriages, see Sleeper-Smith, *Indian Women and French Men*.

98. Wien, "Vie et transfiguration du coureur de bois."

99. Belmessous, "Etre français en Nouvelle-France," 532.

100. This is an insight that has come out of the economic and social histories of New France. See Havard, *Empire et métissages*, 56.

101. Marie Noëlle Bourguet, "Le sauvage, le colon et le paysan," in *Les figures de l'Indien*, ed. Gilles Thérien (Montréal: Typo, 1995), 215.

102. On the emergence of the Métis people during this period, see Jacqueline Peterson, "Many Roads to Red River: Métis Genesis in the Great Lakes Region, 1680–1815," in *The New Peoples: Being and Becoming Métis in North America*, ed. Jacqueline Peterson and Jennifer S. H. Brown (Lincoln: University of Nebraska Press, 1985), 37–72; Olive Patricia Dickason, "From 'One Nation' in the Northeast to 'New Nation' in the Northwest: A Look at the Emergence of the Métis," in *The New Peoples: Being and Becoming Métis in North America*, ed. Jacqueline Peterson and Jennifer S. H. Brown (Lincoln: University of Nebraska Press, 1985), 19–36. Intermarriage was particularly contested in Louisiana, where the introduction of African slaves in larger numbers increased metropolitan anxiety about intermarriage and sexual contact. See Kathleen DuVal, "Indian Intermarriage and Métissage in Colonial Louisiana," *William and Mary Quarterly* 65, no. 2 (2008): 267–304; Jennifer M. Spear, *Race, Sex, and Social Order in Early New Orleans* (Baltimore: Johns Hopkins University Press, 2010), chs. 1, 2; Cécile Vidal, "Caribbean Louisiana: Church, *Métissage*, and the Language of Race in the Mississippi Colony During the French Period," in *Louisiana: Crossroads of the Atlantic World*, ed. Cécile Vidal (Philadelphia: University of Pennsylvania Press, 2014), 125–46.

103. Deslandres, *Croire et faire croire*, 270–73. See also Anderson, *The Betrayal of Faith*.

104. See Greer, *The People of New France*, 82.

105. Numerous scholars of Anglo-American colonies have similarly asserted that the end of the seventeenth century saw an increasingly racialized sense of Native American difference. See Nancy Shoemaker, *A Strange Likeness: Becoming Red and White in Eighteenth-Century North America* (Oxford: Oxford University Press, 2004) for an overview of this process, and, for studies that situate this history in the context of specific moments of colonial violence, see Joyce E. Chaplin, "Natural Philosophy and an Early Racial Idiom in North America: Comparing English and Indian Bodies," *William and Mary Quarterly* 54, no. 1 (1997): 229–52; Jill Lepore, *The Name of War: King Philip's War and the Origins of American Identity* (New York: Knopf, 1998).

106. Havard, "Les forcer"; Cécile Vidal, "Francité et situation coloniale," *Annales. Histoire, Sciences Sociales* 64, no. 5 (2009): 1019–50. This was also an important moment for the articulation of racialized difference throughout the wider French Atlantic world and anticipated eighteenth-century discussions that formalized racial discrimination. See, among others, Guillaume Aubert, "'The Blood of France': Race and Purity of Blood in the French Atlantic World," *William and Mary Quarterly* 61, no. 3 (2004): 439–78; Pierre H. Boulle, "La construction du concept de race dans la France d'ancien régime," *Outre-mers* 89, no. 336–37 (2002): 155–75; Andrew S. Curran, *The Anatomy of Blackness: Science and Slavery in an Age of Enlightenment* (Baltimore: Johns Hopkins University Press, 2011). On the legal implications of these intellectual shifts, see Sue Peabody, *There Are No Slaves in France: The Political Culture of Race and Slavery in the Ancien Régime* (Oxford: Oxford University Press, 1996); Brett Rushforth, *Bonds of Alliance: Indigenous and Atlantic Slaveries in New France* (Chapel Hill: University of North Carolina Press, 2012).

107. Frontenac, "Lettre du gouverneur de Frontenac au ministre (20 Octobre 1691)," *Rapport de l'archiviste de la Province de Québec* (1928): 69.

108. Pierre Margry, *Découvertes et établissements des Français dans l'ouest et dans le sud de l'Amérique Septentrionale, 1614–1754* (Paris: Maisonneuve et cie., 1879–88), 5:158.

109. Warwick states, in fact, that "ambiguity" was a key fact of representations of the *pays d'en haut* more generally. Warwick, "Les 'pays d'en haut,'" 267. See also Dominique Deslandres, "'Et loing de France, en l'une & l'autre mer, Les Fleurs de Liz, tu as fait renommer': Quelques hypothèses touchant la religion, le genre et l'expansion de la souveraineté française en Amérique aux xvie–xviiie siècles," *Revue d'histoire de l'Amérique française* 64, no. 3–4 (2011): 102–3.

110. Donna Haraway uses this term to orient us toward improvised, interspecies dependencies that are "becoming with." Donna J. Haraway, *When Species Meet* (Minneapolis: University of Minnesota Press, 2008), 15–16.

111. "'Mémoire instructif sur le Canada' par l'intendant Champigny," C11A, 11: 262 ANOM. See also Wien, "Vie et transfiguration du coureur de bois."

112. William Bennett Munro, ed., *Documents Relating to the Seigniorial Tenure in Canada, 1598–1854* (Toronto: Champlain Society, 1908), 65.

113. Far more common were generic complaints that not enough land was being cleared by those to whom the crown had given it. The first threat to revoke grants came in 1663, but this was followed up regularly thereafter by proposals to "retrench," "annul," "reunite," or otherwise challenge the "abuse" of *habitants* and seigneurs who failed to adequately clear and cultivate the land. See, for numerous examples of these edicts, Munro, *Documents Relating to the Seigniorial Tenure in Canada*. Serge Courville has suggested that the crown's aim was nothing less than "the total mastery of the territory" claimed by France. Serge Courville, "Espace, territoire et culture en Nouvelle-France: une vision géographique," *Revue de l'histoire de l'Amérique française* 37, no. 3 (1983): 419.

114. Catalogne, "Report on the Seigniories and Settlements."

115. See André Charbonneau, "Sébastien Le Prestre de Vauban: Une vision du développement des colonies," *Cap-aux-Diamants*, no. 92 (2008): 16–20.

116. Sébastien Le Prestre Vauban, *Oisivetés de M. de Vauban* (Paris: J. Corréard, 1843), 4:1. On Vauban's ambivalence, see Miquelon, *New France, 1701–1744*, ch. 1.

117. Vauban, *Oisivetés de M. de Vauban*, 6.

118. Ibid., 8. Vauban was not the first to suggest that this was necessary. See, for an earlier example by Colbert, Jacob Soll, *The Information Master: Jean-Baptiste Colbert's Secret State Intelligence System* (Ann Arbor: University of Michigan Press, 2009), 113–14.

119. This was, then, part of a broader interest in legibility that pervaded the creation of modern states. See James C. Scott, *Seeing Like a State: How Certain Schemes to Improve the Human Condition Have Failed* (New Haven: Yale University Press, 1998).

120. Vauban, *Oisivetés de M. de Vauban*, 8–9.

121. Ibid., 2.

122. Boucher, *Histoire veritable*, 24.

123. Ibid., 23.

124. See, for example, Taylor, *American Colonies*, ch. 16.

125. For a general overview of the western expansion of French colonialism, see Havard, *Empire et métissages*, 102–12. For the study of the adaptation of indigenous communities to these efforts, see Sleeper-Smith, *Indian Women and French Men*, chs. 1, 2. On the westward shift of the fur trade, see Thomas Wien, "Le Pérou éphémère: Termes d'échange et éclatement du commerce Franco-amérindien, 1645–1670," in *Vingt ans apres, Habitants et marchands: Lectures de l'histoire des XVIIe et XVIIIe siècles canadiens* (Montréal: McGill-Queen's University Press, 1998), 160–88.

126. On these critiques, see Robert Michael Morrissey, *Empire by Collaboration: Indians, Colonists, and Governments in Colonial Illinois Country* (Philadelphia: University of Pennsylvania Press, 2015), 43–45.

127. *JR*, 58:97.

128. Jean-François Palomino, "Cartographier la terre des païens: La géographie des missionnaires jésuites en Nouvelle-France au xviie siècle," *Revue de Bibliothéque et Archives nationales du Québec*, no. 4 (2012): 16–17.

129. Margry, *Découvertes et établissements*, 1:122.

130. Riley, *The Once and Future Great Lakes Country*, 46.

131. Margry, *Découvertes et établissements*, 1:126.

132. Boucher, *Histoire veritable*, 19–20. This climatological difference has been extensively noted as a major factor in the agricultural history of the colony. See, for example, Dépatie, "Jardins et vergers à Montréal au XVIIIe siècle"; Wien, "'Les travaux pressants,'" 535–58.

133. Bacqueville de la Potherie, *Histoire de l'Amérique septentrionale* (Paris: J.-L. Nion et F. Didot, 1722), 208, 342.

134. Historian Geoffrey Parker explains that "in the northern hemisphere, 9 of the 14 summers between 1666 and 1679 were either cool or exceptionally cool—harvests in western Europe ripened later in 1675 than in any other year between 1484 and 1879—and climatologists regard the extreme climatic events and disastrous harvests during the 1690s, with average temperatures 1.5 C below those of today, as the 'climax of the Little Ice Age.'" Parker, *Global Crisis*, xxviii.

135. Coates and Degroot, "'Les bois engendrent les frimas et les gelées,'" 206–7. See also Hugo Beltrami and Jean-Claude Mareschal, "Ground Temperature Histories for Central and Eastern Canada from Geothermal Measurements: Little Ice Age Signature," *Geophysical Research Letters* 19, no. 7 (1992): 689–92.

136. Desloges, *Sous les cieux de Québec*, ch. 1.

137. Sam White, *A Cold Welcome: The Little Ice Age and Europe's Encounter with North America* (Cambridge, Mass.: Harvard University Press, 2017), ch. 9.

138. Bronwen Catherine McShea, "Cultivating Empire Through Print: The Jesuit Strategy for New France and the Parisian Relations of 1632 to 1673" (PhD diss., Yale University, 2011), 99.

139. Marie de l'Incarnation, *Lettres de la venerable mere Marie de l'Incarnation premiere superieure des ursulines de la nouvelle France* (Paris: Chez Louis Billaine, 1681), 280.

140. Antoine-Denis Raudot, *Relations par lettres de l'Amérique septentrionale (Années 1709 et 1710)* (Paris: Letouzey et ané, 1904), 12–13.

141. Henri-Louis Duhamel du Monceau, *Avis pour le transport par mer des arbres, des plantes vivaces, des semences, et de diverses autres curiosites d'histoire naturelle* (Paris: L'Imprimerie Royale, 1753), 67.

142. As the historian Yvon Desloges writes, "The winter marked the imaginary of colonists, authors and travelers. It imposed itself." This is one of many influences described in Desloges, *Sous les cieux de Québec*, 4.

143. Coates and Degroot, "'Les bois engendrent les frimas et les gelées,'" 212.

144. Denys Delâge, "L'influence des Amérindiens sur les Canadiens et les Français au temps de la Nouvelle-France," *Lekton* 2, no. 2 (1992): 103–91.

145. Lamontagne, *L'hiver dans la culture québécoise*, 58–59.

146. Pierre Deffontaines, *L'homme et l'hiver au Canada* (Paris: Gallimard, 1957), 144–45.

147. Ibid., ch. 2; Lamontagne, *L'hiver dans la culture québécoise*, 28. In urban areas, fear of fire promoted regulations requiring stone construction. Delaney, "'Le Canada est un país de Bois,'" 138.

148. See, for an overview of French experience in the Caribbean, Philip P. Boucher, *France and the American Tropics to 1700: Tropics of Discontent?* (Baltimore: Johns Hopkins University Press, 2008).

149. Raudot, *Relations par lettres*, 9.

150. Thomas Chapais, *Jean Talon, intendant de la Nouvelle-France (1665–1672)* (Québec: S.-A. Demers, 1904), 415.

151. "Mémoire sur les affaires du Canada: 1 la justice: 2 le commerce: 3 la guerre (Avril 1689)," *Rapport de l'archiviste de la province de Québec* (1923): 7. This was part of a broader debate about the introduction of African slaves and the regulation of enslavement in the colony. See Rushforth, *Bonds of Alliance*, ch. 3.

152. Coates and Degroot, "'Les bois engendrent les frimas et les gelées.'"

153. Talon, "Memoire de Talon sur l'état présent du Canada (1667)," *Rapport de l'archiviste de la province de Québec* (1931): 63.

154. Lahontan, *Oeuvres complètes*, 1019.

155. Ibid., 549, 46.

156. Margry, *Découvertes et établissements*, 4:19.

157. Hennepin, *Description*, 10.

158. Ibid.

159. Richard White, for example, describes a La Salle "who sought to carve out his own fur trade empire in the *pays d'en haut*." White, *The Middle Ground*, 28.

160. Margry, *Découvertes et établissements*, 1:438.

161. Ibid., 2:13.

162. Ibid., 1:465.

163. Ibid., 2:74.

164. Morrissey, "The Power of the Ecotone."

165. Kathleen DuVal, *The Native Ground: Indians and Colonists in the Heart of the Continent* (Philadelphia: University of Pennsylvania Press, 2006), 74–75.

166. Desbarats and Greer, "Où est la Nouvelle-France?"

167. Ingold, *The Perception of the Environment*, 186.

168. Quoted in Claiborne A. Skinner, *The Upper Country: French Enterprise in the Colonial Great Lakes* (Baltimore: Johns Hopkins University Press, 2008), 54.

169. See Ekberg, *French Roots in the Illinois Country*, ch. 1.

170. Ibid., 33.

171. Cécile Vidal, "From Incorporation to Exclusion: Indians, Europeans, and Americans in the Mississippi Valley from 1699 to 1830," in *Empires of the Imagination: Transatlantic Histories of the Louisiana Purchase*, ed. Peter J. Kastor and François Weil (Charlottesville: University of Virginia Press, 2009), 62–63.

172. Robert Michael Morrissey, "Kaskaskia Social Network: Kinship and Assimilation in the French-Illinois Borderlands, 1695–1735," *William and Mary Quarterly* 70, no. 1 (2013): 103–46; White, *Wild Frenchmen and Frenchified Indians*.

CHAPTER 5

1. Cooper, *Inventing the Indigenous*, ch. 2.

2. Steven Shapin and Simon Schaffer, *Leviathan and the Air-Pump: Hobbes, Boyle, and the Experimental Life* (Princeton, N.J.: Princeton University Press, 1985), 507.

3. Boivin, "La flore du Canada en 1708," n.p.

4. Schiebinger, *Plants and Empire*, ch. 5; Müller-Wille, "Walnuts at Hudson Bay," 38–39.

5. Richard White has characterized the entire French presence in North America as Janus-faced. White, *The Middle Ground*, 142.

6. Loïc Charles and Paul Cheney, "The Colonial Machine Dismantled: Knowledge and Empire in the French Atlantic," *Past & Present* 219, no. 1 (2013): 127–63.

7. For more on Michel Sarrazin, see Arthur Vallée, *Un biologiste canadien, Michel Sarrazin, 1659–1735: Sa vie, ses travaux et son temps* (Québec: LS-A. Proulx, 1927).

8. Boivin, "La flore du Canada en 1708," n.p.

9. Ibid.

10. Ibid.

11. In the context of French Atlantic science, the division between amateur and academic science was a matter of both education and recognition by the Académie Royale des Sciences and in many cases was also reflected by inequalities in the state's financial support of colonial botany and plant collection. The contributors I define as amateur naturalists were self-educated and only rarely received financial compensation from the French state, even when it was a matter of compensation for expenses accrued while collecting rare and novel plants. By contrast, while colonial naturalists and members of the Académie Royale des Sciences such as Québec-based Michel Sarrazin did not receive the annual pensions that were provided for Paris-based members, they routinely received "gratifications" that recognized their service to both the crown and Académie with near-annual payments. See, for a discussion of the payments received by Michel Sarrazin, Kathryn A. Young, "Crown Agent-Canadian Correspondent: Michel Sarrazin and the Académie Royale des Sciences, 1697–1734," *French Historical Studies* 18, no. 2 (1993): 427–28. For a more general analysis of the financial support of the Académie Royale des Sciences, see Alice Stroup, *A Company of Scientists: Botany, Patronage, and Community at the Seventeenth-Century Parisian Royal Academy of Sciences* (Berkeley: University of California Press, 1990).

12. Cities such as Paris were fast becoming "nerve centres" in both European and global scientific networks, or "privileged sites for the global articulation of knowledge." Antonella Romano and Stéphane Van Damme, "Science and World Cities: Thinking Urban Knowledge and Science at Large (16th–18th century)," *Itinerario* 33, no. 1 (2009): 84–85. See also Stéphane Van Damme, *Paris, capitale philosophique: De la Fronde à la Révolution* (Paris: Odile Jacob, 2005).

13. For a discussion of polycentricity and its place in reconstructions of colonial science in both the Atlantic and world history, see Fa-ti Fan, "Science in Cultural Borderlands: Methodological Reflections on the Study of Science, European Imperialism, and Cultural Encounter," *East Asian Science, Technology and Society: An International Journal* 1, no. 2 (2007): 212–31. For a discussion of biocontact zones, see Chapter 3.

14. Numerous scholars have studied the close relationship between science and state in seventeenth- and eighteenth-century France. See, for example, Charles Coulton Gillespie, *Science and Polity in France: The End of the Old Regime* (Princeton, N.J.: Princeton University Press, 2004); James E. McClellan and François Regourd, *The Colonial Machine: French Science and Overseas Expansion in the Old Regime* (Turnhout: Brepols, 2011); Emma C. Spary, *Utopia's Garden: French Natural History from Old Regime to Revolution* (Chicago: University of Chicago Press, 2000); idem, "'Peaches which the patriarchs lacked': Natural History, Natural Resources, and the Natural Economy in France," *History of Political Economy* 35, no. 5 (2003): 14–41.

15. Sarrazin particularly benefited from the patronage of intendants Michel Bégon and Antoine-Denis Raudot. See Young, "Crown Agent–Canadian Correspondent," 420–22. The superior council (sovereign council before 1703) was composed of the governor, the intendant, the bishop, a magistrate, a clerk, and other prominent subjects. It functioned primarily as a judicial body and a court of appeal. Havard and Vidal, *Histoire de l'Amérique française*, 159–61.

16. Vallée, *Un biologiste canadien*, 42.

17. On Sarrazin's first botanizing trips, see Rousseau, "Michel Sarrazin, Jean-François Gaulthier et l'étude prélinnéenne de la flore canadienne," in *Les botanistes français en Amérique du Nord avant 1850* (Paris: Centre national de la recherche scientifique, 1957), 153–54.

18. Gilles-Antoine Langlois, "Deux fondations scientifiques à la Nouvelle-Orléans (1728–30): La connaissance à l'épreuve de la réalité coloniale," *French Colonial History* 4, no. 1 (2003): 99–115. Although not explicitly described as a model, it seems inevitable that the example of state-supported science in the Iberian Atlantic inspired French practice. See, for an overview of this subject, Daniela Bleichmar, Paula de Vos, Kristin Huffine, and Kevin Sheehan, eds., *Science in the Spanish and Portuguese Empires, 1500–1800* (Stanford, Calif.: Stanford University Press, 2009).

19. In botanists such as New France's Michel Sarrazin and Jean-François Gaultier or Louisiana's Jean and Louis Prat and Alexandre Vielle, historians James McClellan and François Regourd have seen the birth of what they term "the colonial machine," a long-term process by which scientific activity was brought under the supervision of the Parisian academies and was integrated into the administrative apparatus of the French Atlantic empire. They have more specifically traced the origins of this machine to larger efforts to centralize the production and circulation of knowledge in France during the rule of Louis XIV and the ascendance of ministers such as Jean-Baptiste Colbert. McClellan and Regourd, *The Colonial Machine*, 43–44; McClellan has more fully traced the growth of the colonial machine following the reorganization of the French Atlantic empire after the Seven Years' War. James E. McClellan, *Colonialism and Science: Saint Domingue in the Old Regime* (Chicago: University of Chicago Press, 1992).

20. Kenneth J. Banks, *Chasing Empire Across the Sea: Communications and the State in the French Atlantic, 1713–1763* (Kingston: McGill-Queen's University Press, 2002), 23.

21. Pritchard, *In Search of Empire*, 234–41.

22. Roger Hahn, *The Anatomy of a Scientific Institution: The Paris Academy of Sciences, 1666–1803* (Berkeley: University of California Press, 1971), 46–47, 58–60.

23. Charles Gillespie has written that the Académie "exhibited in miniature the structural characteristics of the regime that sustained [it]: monarchical, hierarchical, prescriptive, and privileged." Gillespie, *Science and Polity in France*, 81.

24. Hahn, *The Anatomy of a Scientific Institution*, 63, 73–75.

25. David L. Sturdy, *Science and Social Status: The Members of the Académie Royale des Sciences, 1666–1760* (Woodbridge, UK: Boydell Press, 1995), 286. For an overview of the history of the reorganization of the Académie, see Marie-Jeanne Tits-Dieuaide, "Les savants, la société et l'État: À propos du 'renouvellement' de l'Académie royale des sciences (1699)," *Journal des savants* (1998): 79–114.

26. James E. McClellan, "L'Académie royale des Sciences (1666–1793)," in *Lieux de Savoir: Espaces et communautés*, ed. Christian Jacob (Paris: Albin Michel, 2007), 719–22.

27. Stroup, *A Company of Scientists*, 14.

28. Ibid.

29. James E. McClellan, "The Académie Royale des Sciences, 1699–1793: A Statistical Portrait," *Isis* 72, no. 4 (1981): 544. On Jesuit members, see Florence C. Hsia, "Jesuits, Jupiter's Satellites, and the Académie Royale des Sciences," in *The Jesuits: Cultures, Sciences, and the Arts, 1540–1773*, ed. John W. O'Malley et al. (Toronto: University of Toronto Press, 1999), 241–57.

30. Sturdy, *Science and Social Status*, 284.

31. Hsia, *Sojourners in a Strange Land*, 118–19.

32. Sturdy, *Science and Social Status*, 286.

33. Ibid.

34. McClellan and Regourd, *The Colonial Machine*, 72–78.

35. See, for example, the work of Florence Hsia on the recruitment of observers by Cassini and other astronomers of the Académie. Hsia, *Sojourners in a Strange Land*.

36. Bégon, cousin to Colbert and a botanist himself, was successively intendant in the French Caribbean and in the French ports of Rochefort and La Rochelle. His son Michel Bégon de la Picardière was named intendant of New France in 1710. On this family, see Yvonne Bezard, *Fonctionnaires maritimes et coloniaux sous Louis XIV: Les Bégon* (Paris: Albin Michel, 1932).

37. Spary, *Utopia's Garden*, ch. 2.

38. This is the image of the "colonial machine" put forward in McClellan and Regourd, *The Colonial Machine*.

39. Roland Lamontagne, "L'influence de Maurepas sur les sciences: Le botaniste Jean Prat à La Nouvelle-Orléans, 1735–1746," *Revue d'histoire des sciences* 49, no. 1 (1996): 118.

40. Sarrazin to Réaumur, October 10, 1726, in Vallée, *Un biologiste canadien*, 219.

41. Joseph Pitton de Tournefort, *Elemens de botanique, ou, Méthode pour connoître les plantes* (Paris: Chez Pierre Bernuset, 1797), 49. See also Lorraine Daston, "At the Center and the Periphery: Joseph Pitton de Tournefort Botanizes in Crete," in *Relocating the History of Science: Essays in Honor of Kostas Gavroglu*, ed. Theodore Arabatzis, Jürgen Renn, and Ana Simões (Cham: Springer International Publishing, 2015).

42. Sharon Kettering, *Patronage in Sixteenth- and Seventeenth-Century France* (Farnham, UK: Ashgate, 2002), 14.

43. Duhamel du Monceau, *Traité des arbres et arbustes*, v–vi.

44. Ibid., 31.

45. Ibid., 56.

46. Ibid., 108.

47. Sébastien Vaillant, *Botanicon Parisiense, ou, Denombrement par ordre alphabetique des plantes qui se trouvent aux environs de Paris* (Leiden: J. & H. Verbeek, 1727), n.p.

48. Staffan Müller-Wille, "Joining Lapland and the Topinambes in Flourishing Holland: Center and Periphery in Linnaean Botany," *Science in Context* 16, no. 4 (2003): 461.

49. Steven J. Harris, "Long-Distance Corporations, Big Sciences, and the Geography of Knowledge," *Configurations* 6, no. 2 (1998): 294.

50. Van Damme, *Paris, capitale philosophique*. For an example of how this process worked in late eighteenth-century England, see David Philip Miller, "Joseph Banks, Empire and 'Centers of Calculation' in Late Hanoverian London," in *Visions of Empire: Voyages, Botany, and Representation of Nature*, ed. David Philip Miller and Peter Hans Reill (Cambridge: Cambridge University Press, 1995), 21–37.

51. Tournefort, *Elemens de botanique*, 47.

52. Staffan Müller-Wille, "Nature as a Marketplace: The Political Economy of Linnaean Botany," *History of Political Economy* 35, Annual Supplement (2003): 160.

53. Lamontagne, "L'influence de Maurepas sur les sciences," 117.

54. Duhamel's guide was published several times; see Duhamel du Monceau, *Avis pour le transport par mer.* I was able to find reference to a guide written for the collectors of Tournefort only in passing. See "Copie d'une lettre ecritte à Pensacola le 15 Janvier 1714," in *Memoranda on French Colonies in America, including Canada, Louisiana, and the Caribbean [1702–1750],* p. 18, MS 293, Edward E. Ayers Collection, Newberry Library, Chicago.

55. This is paraphrased from Daston and Galison, *Objectivity,* 38–39.

56. Lorraine Daston, "Attention and the Values of Nature in the Enlightenment," in *The Moral Authority of Nature,* ed. Lorraine Daston and Fernando Vidal (Chicago: University of Chicago Press, 2004).

57. Tournefort, *Elemens de botanique,* 49. For a detailed examination of botanical study during this period in France, see Williams, *Botanophilia in Eighteenth-Century France: The Spirit of the Enlightenment* (Dordrecht: Kluwer Academic Publishers, 2001).

58. "Maniere de faire du Bray," Group no. 20, Duhamel du Monceau Papers, American Philosophical Society, Philadelphia.

59. "Observations botaniques," *HMARS* (1730): 65–66

60. Jean-François Gaultier, "Description de plusieurs plantes du Canada par M Gaultier," pp. 10–11, Cote: P91, D3, Fonds Jean-François Gaultier, Bibliothèque et Archive nationale de Québec, Québec.

61. Ibid.

62. Vaillant, *Botanicon Parisiense,* n.p.

63. Ogilvie, *The Science of Describing,* 214.

64. For an overview of Kalm's expedition, see Müller-Wille, "Walnuts at Hudson Bay," 34–48.

65. I have consulted a recently edited and annotated version that focuses on Kalm's travels in New France. See Pehr Kalm, *Voyage de Pehr Kalm au Canada en 1749* (Montréal: P. Tisseyre, 1977).

66. For a list of the plants requested by Linnaeus, see ibid., 533.

67. Ibid., 365.

68. Ibid., 233; Müller-Wille, "Walnuts at Hudson Bay," 44–48.

69. Kalm, *Voyage de Pehr Kalm,* 247.

70. The commercial trade in plants is beyond the scope of this chapter but has been excellently studied by Sarah Easterby-Smith. See, for example, Sarah Easterby-Smith, "Selling Beautiful Knowledge: Amateurship, Botany and the Market-Place in Late Eighteenth-Century France," *Journal for Eighteenth-Century Studies* 36, no. 4 (2013): 531–43; idem, *Cultivating Comerce: Cultures of Botany in Britain and France, 1760–1815* (Cambridge: Cambridge University Press, 2018).

71. For this sort of hagiography, see the work on Sarrazin done almost a century ago, notably Vallée, *Un biologiste canadien.* See also J. C. K. Laflamme, "Michel Sarrazin: Matériaux pour servir à l'histoire de la science en Canada," *Transactions of the Royal Society of Canada* 5, no. 4 (1887): 1–23.

72. In that way, naturalists such as Sarrazin were not that dissimilar from other figures in the French Atlantic world who sought to expand and maintain communication networks. Kenneth Banks writes, for example, of "the challenge of trying to absorb, comprehend, evaluate,

and coordinate a very complex number of tasks in a wide variety of climates across a vast ocean." Banks, *Chasing Empire Across the Sea*, 5.

73. Bauer, *The Cultural Geography of Colonial American Literatures*, 4.

74. See Chambers and Gillespie, "Locality in the History of Science: Colonial Science, Technoscience, and Indigenous Knowledge," *Osiris* 15 (2000): 221–40.

75. Frederick Cooper and Laura Ann Stoler, "Between Metropole and Colony: Rethinking a Research Agenda," in *Tensions of Empire: Colonial Cultures in a Bourgeois World*, ed. Frederick Cooper and Laura Ann Stoler (Berkeley: University of California Press, 1997), 24.

76. "Le comte de Pontchartrain à M. Raudot," June 29, 1707, C11G, 2:104–104v, ANOM

77. Young, "Crown Agent-Canadian Correspondent," 425.

78. Steven Shapin, *A Social History of Truth: Civility and Science in Seventeenth-Century England* (Chicago: University of Chicago Press, 1994), ch. 8.

79. "Lettre de Beauharnois au ministre," October 17, 1736, C11A, 65:140, ANOM.

80. "Lettre de Chaussegros de Léry fils au ministre," November 2, 1749, C11A, 94:67–67v, ANOM. This was similar to realities at the Jardin du Roi, where botanists received numerous gifts of specimens that had originally been sent to the king. Spary, *Utopia's Garden*, ch. 2.

81. Kalm, *Voyage de Pehr Kalm*, 314–15, 405–8. The Marquis de la Galissonnière was also a respected naturalist in his own right, best remembered for his work with Duhamel du Monceau. Several examples of his work can be seen in Duhamel du Monceau's papers at the American Philosophical Society. For broader studies of Galissonnière's scientific activities, see Roland Lamontagne, *La Galissonnière et le Canada* (Montréal: Presses de l'Université de Montréal, 1962), 68–88.

82. Kalm, *Voyage de Pehr Kalm*, 409–10.

83. On "obligatory points of passage," see John Law, "Technology, Closure and Heterogenous Engineering: The Case of Portuguese Expansion," in *The Social Construction of Technological Systems*, ed. Wiebe Bijker, Trevor Pinch, and Thomas P. Hughes (Cambridge, Mass.: MIT Press, 1987), 111–34. On the transport of plants by sea, see Christopher M. Parsons and Kathleen S. Murphy, "Ecosystems Under Sail: Specimen Transport in the Eighteenth-Century French and British Atlantics," *Early American Studies* 10, no. 3 (2012): 503–29; Yannick Romieux, "Le transport maritime des plantes au XVIIIe siècle," *Revue d'histoire de la pharmacie* 92, no. 343 (2004): 405–18.

84. Rousseau, "Michel Sarrazin, Jean-François Gaulthier et l'étude prélinnéenne de la flore canadienne," 153.

85. For a discussion of disinterestedness and its role in authenticating knowledge in early modern science, see Shapin, *A Social History of Truth*. For a discussion of colonists who sought to participate in English Atlantic scientific networks, see Parrish, *American Curiosity*, ch. 3.

86. "Délibération du Conseil de Marine sur une lettre de Vaudreuil et Bégon," January 1717, C11A, 37:31, ANOM.

87. "Lettre du médecin Jean-François Gaultier au minister," November 1, 1749, C11A, 94:45v, ANOM.

88. Gaultier to Réaumur, October 30, 1750, in Vallée, "Cinq lettres inédites de Jean François Gaultier à M. de Rhéaumur de l'Académie des sciences," *Mémoire de la Société royale du Canada* 24 (1930): 36.

89. Galissonnière to Duhamel du Monceau, October 28, 1748, Group no. 15, Duhamel du Monceau Papers, American Philosophical Society, Philadelphia.

90. Gaultier to Réaumur, October 28, 1753, in Vallée, "Cinq lettres inédites," 39.

91. This is part of a broader "frontier exchange economy." See Usner, *Indians, Settlers, & Slaves in a Frontier Exchange Economy*.

92. Kalm, *Voyage de Pehr Kalm*, 3.

93. See, for example, ibid., 217 where he provides the common French names for Canadian maples. Aboriginal people do, on occasion, appear as guides, but their contribution to his botanical study, while mentioned from time to time, is rare. See ibid., 250.

94. Ibid., 7, 359, 389.

95. Ibid., 470.

96. Boivin, "La flore du Canada en 1708," n.p.

97. Ibid.

98. Jean-François Gaultier, "Description de plusieurs plantes du Canada par M Gaultier," pp. 5–9, Cote: P91, D3, Fonds Jean-François Gaultier, Bibliothèque et Archive nationale de Québec, Québec.

99. Boivin, "La flore du Canada en 1708," n.p.

100. Gaultier, "Description," 29.

101. Kathleen Murphy suggests that colonial naturalists in the Anglo-Atlantic world treated enslaved and indigenous people in a similar fashion, attributing their knowledges to a collective identity. Murphy, "Translating the Vernacular," 33.

102. See, for examples, "Lettre non signee," October 18, 1727, C11A, 49:519, ANOM; "Lettre de Hocquart au minister," October 23, 1735, C11A, 64:147–49, ANOM; "Lettre de Beauharnois au ministre," October 17, 1736, C11A, 65:140–141v, ANOM; "Lettre de Hocquart au ministre," October 19, 1738, C11A, 70:113–114v, ANOM; "Lettre de Hocquart au ministre," October 26, 1738, C11A, 70:124–26, ANOM.

103. See "La Croix, Hubert-Joseph de," *Dictionary of Canadian Biography Online*, http://www.biographi.ca/en/bio.php?id_nbr=1459.

104. "Lettre de Hocquart au ministre," October 19, 1738, C11A, 70:113, ANOM; "Lettre de Hocquart au ministre," November 8, 1740, C11A, 73:416, ANOM.

105. See Parsons and Murphy, "Ecosystems Under Sail."

106. Lamontagne, "L'influence de Maurepas sur les sciences," 119.

107. Parrish, *American Curiosity*, 114.

108. "Lettre de Hocquart au ministre," October 26, 1738, C11A, 70:125, ANOM; "Mémoire des graines de différentes plantes et des autres morceaux d'histoire naturelle," 1745, C11A, 72:174, ANOM.

109. "État de la dépense que moi Gosselin," October 26, 1738, C11A, 70:127, ANOM.

110. Ibid. There is little evidence to study whether aboriginal peoples were routinely paid for botanical specimens. Kathleen Murphy, however, has recently suggested that this was common practice in British North America. Murphy, "Translating the Vernacular," 35–36.

111. See "Gosselin, Jean-Baptiste," *Dictionary of Canadian Biography Online*, http://www.biographi.ca/en/bio.php?id_nbr=1378.

112. A thorough analysis of Lafitau's discovery of ginseng is presented in Chapter 6.

113. "Le Genseng, Plante si précieuse à la Chine, découverte dans le Canada," *Mémoires pour l'Histoire des Sciences & des Beaux-Arts* (1717): 121–24.

114. Joseph-François Lafitau, *Mémoire presenté à son altesse royale Monseigneur le Duc d'Orleans, regent du royaume de France, concernant la précieuse plante du gin-seng de Tartarie, découverte en Canada* (Paris: Chez Joseph Mongé, 1718).

115. On "biocontact zones," see Chapter 3.

116. See Schiebinger, *Plants and Empire*.

117. Vallée, "Memoire sur les plantes qui sont dans le caise B," Laboratoire de phanérogamie, Muséum national d'Histoire naturelle, Paris. The dating was provided by researchers at the Laboratoire de phanérogamie and remains an estimate. While the text rests anonymous at the MNHN today, a list of possible authors was produced in consultation with Dr. Kenneth Donovan, historian at the Fortress of Louisbourg—Louisbourg was, after all, not so big a settlement in 1725 that there were too many possibilities. Working from this list, I compared handwriting samples and searched colonial archives for circumstantial evidence that might suggest the identity of the author. A parallel reading of archival documents demonstrated that François-Madeleine Vallée was known to colonial and metropolitan authorities for his interest in botany and that he later claimed to have some expertise in the subject. For Donovan's own analysis of botany and horticulture at Louisbourg, see Kenneth Donovan, "Imposing Discipline upon Nature: Gardens, Agriculture and Animal Husbandry in Cape Breton, 1713–1758," *Material Culture Review/ Revue de la culture matérielle* 64 (2006): 20–37. For more on Vallée, see "Vallée, François-Madeleine," *Dictionary of Canadian Biography Online*, http://www.biographi.ca/en/bio.php?id_nbr=1696.

118. Vallée, "Memoire sur les plantes," n.p.

119. There is some evidence that these taxonomic systems were known in the colonies. See, for example, "Relation de la Louisiane, par Le Maire," MS 948, Bibliothèque centrale du Muséum national d'Histoire naturelle, Paris.

120. Vallée, "Memoire sur les plantes," n.p.

121. See Chapter 6.

122. "*Panax quinquefolium*," Herbier d'Antoine Laurent de Jussieu, Laboratoire de phanérogamie, Muséum national d'Histoire naturelle, Paris.

123. See Antoine de Jussieu, "Histoire du gin-sem et ses qualités," p. 1, MS 1151, Bibliothèque centrale du Muséum national d'Histoire naturelle, Paris; Sébastien Vaillant, *Discours sur la structure des fleurs, leurs différences et l'usage de leurs parties* (Leide: Chez Pierre Vander, 1718).

124. Gaultier, "Description," 97.

125. Greer, "The Exchange of Medical Knowledge Between Natives and Jesuits in New France," 135–36; Havard, *Empire et métissages*, 607–8.

126. Alexandre Vielle, "Description de l'arbrisseau qui porte la cire," p. 3, MS 196, Bibliothèque centrale du Muséum national d'Histoire naturelle.

127. The importance of indigenous knowledges to Jesuit accounts of ginseng and other American plants and more general discussion of the mediating effects of indigenous knowledge are studied in Chapters 3 and 6.

CHAPTER 6

1. Académie Royale des Sciences, *Procès-verbaux* 36 (1717): 11.

2. Ibid. There was no standard spelling for "ginseng" in the eighteenth century. Rather than standardize the spelling, or including [*sic*] in each instance where the spelling varies, I have left the original spelling of the word unaltered.

3. Lafitau, *Mémoire*, 14.

4. Ibid., 61.

5. Bauer, *The Cultural Geography of Colonial American Literatures*, 4.

6. Bignon was the nephew of minister of the marine Louis Phélypeaux, Comte de Pontchartrain. See Stroup, *A Company of Scientists*.

7. Sarrazin to Bignon, November 5, 1717, in Vallée, *Un biologiste canadien*, 216.

8. Ibid.

9. Boivin, "La flore du Canada en 1708," 281. *Ornithogalum* is a genus of flowering plants native to Europe and Africa commonly called "Star-of-Bethlehem."

10. The *Journal de Trévoux* sought a popular, although still educated, audience and was intended to compete for the attention of readers who might otherwise turn to Protestant journals for coverage of recent scientific discoveries and publications. See Stéphane Van Damme, "Education, Sociability and Written Culture: The Case of the Society of Jesus in France," Blumenthal Lectures, Cornell University, October 2002, Les Dossiers du Grihl, http://dossiersgrihl.revues.org/752.

11. For a recent overview of Jesuit natural history in North and South America, see the essays in Domingo Ledezma and Luis Millones Figueroa, eds., *El saber de los jesuitas, historias naturales y el Neuvo Mundo* (Frankfurt: Vervuert-Iberoamericana, 2005); Steven J. Harris, "Jesuit Scientific Activity in the Overseas Missions, 1540–1773," *Isis* 96, no. 1 (2005): 71–79. Jesuit natural histories of Spanish American colonies are better known than their French counterparts. See, for instance, Sabine Anagnostou, "Jesuit Missionaries in Spanish America and the Transfer of Medical-Pharmaceutical Knowledge," *Archives internationales d'Histoire des Sciences* 52 (2002): 176–97; Miguel de Asúa, "'Names which he loved, and things well worthy to be known': Eighteenth-Century Jesuit Natural Histories of Paraquaria and Río de la Plata," *Science in Context* 21, no. 1 (2008): 39–72.

12. This is discussed further in Chapter 3.

13. Both the *Relations* and the *Lettres édifiantes et curieuses* were serial publications that featured writing from Jesuit missionaries overseas. The *Lettres édifiantes et curieuses* began publishing in 1702. Accounts of American missions were assembled with accounts of the society's global mission, but the volumes focused particular attention of the missions to Southeast Asia. For a recent discussion of the American content in the *Lettres édifiantes et curieuses*, see Catherine Desbarats, ed., *Lettres édifiantes et curieuses écrites par des missionnaires de la Compagnie de Jésus* (Montréal: Boréal, 2006).

14. There are several such similarities. Both the article in the *Journal de Trévoux* and Lafitau's *Mémoire* contain the phrase "sans une communication d'idées, & par consequent de personnes," for example, and describe a remedy used by an indigenous woman who was "tormented" by a fever; see Lafitau, "Le Genseng, Plante si précieuse," 122; and idem, *Mémoire*, 15, 16.

15. Pierre Jartoux, "Lettre du Père Jartoux au Père Procureur général des missions des Indes et de la Chine," in *Lettres édifiantes et curieuses concernant l'Asie, l'Afrique et l'Amérique*, ed. Louise-Aimé Martin (Paris: Société du Panthéon littéraire, 1843), 3:183–87.

16. Pierre Jartoux identified the location as a village four leagues from Korea (ibid., 183). For a brief overview of the ten-year mapping project directed by the Jesuits, see Joanna Waley-Cohen, *The Sextants of Beijing: Global Currents in Chinese History* (New York: W. W. Norton, 1999), 112–14. For a broader introduction to Jesuit scientific activity in their Chinese missions, see Hsia, *Sojourners in a Strange Land*.

17. Earlier images had circulated in the texts of seventeenth-century naturalists such as Nehemiah Grew and John Ray. See John H. Appleby, "Ginseng and the Royal Society," *Notes and Records of the Royal Society of London* 37, no. 2 (1983): 121–45. For an excellent overview of the role of Jesuits in disseminating information about ginseng in the seventeenth century, see Michael Block, "New England Merchants, the China Trade, and the Origins of California" (PhD diss., University of Southern California, 2011), 70–72.

18. This botanical ambiguity is not unlike the history of other early modern commodities, such as "China Root." Anna E. Winterbottom, "Of the China Root: A Case Study of the Early Modern Circulation of Materia Medica," *Social History of Medicine* 28, no. 1 (2015): 22–44.

19. In 1713, a shorter version was translated into English and published in the Royal Society of London's journal, the *Philosophical Transactions*. Pierre Jartoux, "The Description of a Tartarian Plant, Call'd Gin-Seng: With an Account of Its Virtues. In a Letter from Father Jartoux, to the Procurator General of the Missions of India and China. Taken from the Tenth Volume of Letters of the Missionary Jesuits, Printed at Paris in Octavo, 1713," *Philosophical Transactions* 28 (1713): 237–47.

20. Jartoux, "Lettre," 184.

21. Lafitau clearly profited from an apprenticeship with Jesuit missionaries who had spent decades in the missions of French North America while he was at Québec, Sillery, and Kahnawake. See Chapter 3.

22. Lafitau, "Le Genseng, Plante si précieuse," 123.

23. Ibid., 122.

24. Ibid.

25. Ibid.

26. Andreas Motsch has recently characterized this as an epistemological clash between a culture of curiosity and antiquarianism and what we now recognize as scientific modes of thought. Andreas Motsch, "Le ginseng d'Amérique: Un lien entre les deux Indes, entre curiosité et science," *Etudes Epistémè: Revue de littérature et de civilisation (XVIe–XVIIIe siècles)* 26 (2014), http://episteme.revues.org/331.

27. Académie Royale des Sciences, *Procès-verbaux* 36 (1717): 11.

28. Ibid., 310.

29. Wolfgang Muntschick, "Plants That Carry His Name: Engelbert Kaempfer's Study of the Japanese Flora," in *The Furthest Goal: Engelbert Kaempfer's Encounter with Tokugawa Japan*, ed. Beatrice M. Bodart-Bailey and Derek Massarella (Kent: Japan Library, 1995), 81. See also Engelbert Kaempfer, *Kaempfer's Japan: Tokugawa Culture Observed*, ed. Beatrice M. Bodart-Bailey (Honolulu: University of Hawaii Press, 1999).

30. Muntschick, "Plants That Carry His Name," 79–80, 89. Modern botanists and historians believe he had confused ginseng with the Asian perennial "sugar root" (*Sium sisarum L.*); see ibid., 89.

31. Ibid., 309–10.

32. Lafitau, "Le Genseng, Plante si précieuse," 123–24.

33. Académie Royale des Sciences, *Procès-verbaux* 36 (1717): 310.

34. Ibid.

35. Like Danty d'Isnard, Vaillant had been elected a member of the Académie in 1716, but he had already been employed at the Jardin du Roi for over two decades where he had worked under the supervision of and in collaboration with the celebrated botanist Joseph Pitton de Tournefort. Williams, *Botanophilia in Eighteenth-Century France*, 12–13.

36. Académie Royale des Sciences, *Procès-verbaux* 37 (1718): 65. The *Discours* was itself a copy of a lecture that had been "pronounced at the opening of the Jardin du Roi de Paris, the 10th day of the month of June 1717." Vaillant, *Discours sur la structure des fleurs*.

37. Roger Williams makes the case that Vaillant's *Discours* is a landmark in botanical history that has been overlooked in the centuries since because of Vaillant's own fall from grace

within the Académie that followed his challenge to the intellectual legacy of Joseph Pitton de Tournefort. See Williams, *Botanophilia in Eighteenth-Century France*, 9–18.

38. Vaillant, *Discours sur la structure des fleurs*, 40.

39. See Atran, *Cognitive Foundations of Natural History*, 151–81.

40. This is what Staffan Müller-Wille has called the "taxonomic gaze": Müller-Wille, "Walnuts at Hudson Bay," 37–38.

41. Vaillant, *Discours sur la structure des fleurs*, 40.

42. Ibid.

43. Académie Royale des Sciences, *Procès-verbaux* 37 (1718): 32–33.

44. Although Cornut's 1635 work was ostensibly the first of this type, it was not entirely focused on Canadian plants. See Pringle, "How 'Canadian' Is Cornut's *Canadensium Plantarum Historia*?"

45. Lafitau, "Le Ginseng, Plante si précieuse," 121.

46. Lafitau, *Mémoire*, 6–7.

47. Ibid., 7.

48. Clossey, *Salvation and Globalization in the Early Jesuit Missions*; Deslandres, *Croire et faire croire*.

49. Lafitau, *Mémoire*, 13. He likely underestimated this exchange. See Fenton, *Contacts Between Iroquois Herbalism and Colonial Medicine*.

50. This has been a point of confusion in recent literature. Steven Harris suggested that it was a female Mohawk healer who found the plant for Lafitau. Harris, "Jesuit Scientific Activity in the Overseas Missions," 77–78. Sandra Harding, who cites Harris, similarly used Lafitau as evidence of the contribution of indigenous women to colonial science. Sandra Harding, "Postcolonial and Feminist Philosophies of Science and Technology: Convergences and Dissonances," *Postcolonial Studies* 12, no. 4 (2009): 408. For an effort to study Iroquoian medicine at this time, see Gérard Fortin, "La pharmacopée traditionnelle des Iroquois: Une étude ethnohistorique," *Anthropologie et Sociétés* 2, no. 3 (1978): 117–38.

51. James B. McGraw, Suzanne M. Sanders, and Martha Van der Voort, "Distribution and Abundance of Hydrastis canadensis L. (Ranunculaceae) and Panax quinquefolius L. (Araliaceae) in the Central Appalachian Region," *Journal of the Torrey Botanical Society* 130, no. 2 (2003): 62.

52. Roger C. Anderson et al., "The Ecology and Biology of Panax quinquefolium L. (Araliaceae) in Illinois," *American Midland Naturalist* 129, no. 2 (1993): 368.

53. Lafitau, *Mémoire*, 67.

54. These are plants at their reproductive height, producing considerable amounts of fruit and seeds compared to younger plants that will often not reproduce. The average age of reproductive plants in one representative population was observed at 7.1 ± 2.7, 7.5 ± 2.7, and 8.4 ± 2.7. Walter H. Lewis and Vincent E. Zenger, "Population Dynamics of the American Ginseng Panax quinquefolium (Araliaceae)," *American Journal of Botany* 69, no. 9 (1982): 1486. Lafitau's description of fruit that usually bear two seeds conforms to recent studies that have suggested that between two-thirds and three-fourths of fruit are of this type. Anderson et al., "The Ecology and Biology of Panax quinquefolium L. (Araliaceae) in Illinois," 370.

55. While ginseng populations can vary greatly across time and space, researchers examining the sustainability of ginseng populations outside of Montréal in the 1980s found between 60 and 132 specimens in each cluster. These researchers were nonetheless concerned for the plant's viability in Québec, arguing that it required "stable habitats" and could even then support only low levels of harvesting. Danielle Charron and Daniel Gagnon, "The Demography

of Northern Populations of Panax Quinquefolium (American Ginseng)," *Journal of Ecology* 79, no. 2 (1991): 431–32, 37.

56. Lafitau, *Mémoire*, 67.

57. Anderson et al., "The Ecology and Biology of Panax quinquefolium L. (Araliaceae) in Illinois," 360. Habitat destruction is regularly cited as one of the leading causes of the plant's contemporary scarcity. See Charron and Gagnon, "The Demography of Northern Populations of Panax Quinquefolium (American Ginseng)."

58. James B. McGraw and Mary Ann Furedi, "Deer Browsing and Population Viability of a Forest Understory Plant," *Science* 307, no. 5711 (2005): 920–22; James B. McGraw et al., "Ecology and Conservation of Ginseng (Panax quinquefolius) in a Changing World," *Annals of the New York Academy of Sciences* 1286, no. 1 (2013): 62–91.

59. Lafitau, *Mémoire*, 14.

60. Ibid., 13.

61. Ibid., 14.

62. Ibid., 14–15.

63. Ibid., 48.

64. Ibid., 50.

65. Ibid., 16–17.

66. Ibid., 17.

67. Ibid.

68. Ibid.

69. Ibid. Lafitau was not the first Jesuit to suggest a physical connection between Asia and North America. José de Acosta offered a similar theory in his *Natural and Moral History of the Indies* (1590). Anthony Pagden, *The Fall of Natural Man: The American Indian and the Origins of Comparative Ethnology* (Cambridge: Cambridge University Press, 1982), 195.

70. Vaillant, *Discours sur la structure des fleurs*, 2. Essential qualities were considered those least likely to change from place to place and were considered a more stable basis of classification than those parts of the plants that were accidental and that seemed the product of chance and local ecological conditions. See Williams, *Botanophilia in Eighteenth-Century France*, 9–17.

71. Lafitau, *Mémoire*, 32, 61.

72. Lafitau explained that he chose the name *Aureliana Canadensis* in honor of the Duc d'Orléans. Ibid., 86–87.

73. An overview of this *mémoire* and his efforts to aid the Haudenosaunee can be found in *JR*, 67:38.

74. Brian L. Evans, "Ginseng: Root of Chinese-Canadian Relations," *Canadian Historical Review* 66, no. 1 (1985): 10.

75. Arnold H. Rowbotham, "The Jesuit Figurists and Eighteenth-Century Religious Thought," *Journal of the History of Ideas* 17, no. 4 (1956): 476.

76. Ibid., 476–79.

77. Andreas Motsch, *Lafitau et l'émergence du discours ethnographique* (Sillery: Septentrion, 2001), 39. For other studies of Lafitau's method in his *Moeurs*, see William N. Fenton and Elizabeth L. Moore, introduction to *Customs of the American Indians Compared with the Customs of Primitive Times*, ed. and trans. William N. Fenton and Elizabeth L. Moore (Toronto: Champlain Society, 1974–77); Pagden, *The Fall of Natural Man*, ch. 8.

78. Motsch, *Lafitau et l'émergence du discours ethnographique*, ch. 1.

79. For Lafitau "the question of identity," writes Andreas Motsch, "whether that of human beings or that of ginseng, returns to the unity of creation and the integrity of Providence."

Motsch, "Le ginseng d'Amérique," 5. The text therefore spoke directly to many of the concerns that were animating contemporary debates about the origins and history of humanity: those scholars who preceded the *Mémoire* such as Isaac La Peyrère, who had theorized a limited deluge and multiple creations, and those who followed such as Carolus Linnaeus or the Comte de Buffon Georges-Louis Leclerc, who debated the origins and dispersal of the world's life forms. See John C. Briggs and Christopher J. Humphries, "Early Classics," in *Foundations of Biogeography: Classic Papers with Commentaries*, ed. Mark V. Lomolino, Dov F. Sax, and James H. Brown (Chicago: University of Chicago Press, 2004); Curran, *The Anatomy of Blackness*, 11–17.

80. Lafitau, *Mémoire*, 59–60.

81. Ibid., 11.

82. Ibid., 58.

83. Ibid., 74–79.

84. Ibid., 82–83.

85. Ibid., 49.

86. J. B. Du Halde, *Description géographique, historique, chronologique, politique, et physique de l'empire de la Chine et de la Tartarie chinoise* (Paris: P. G. Le Mercier, 1735), 3:460–74.

87. Lafitau, *Mémoire*, 50. In subjecting the Iroquoian *sauvagesse* to this medical experiment, Lafitau joined many other colonial naturalists who experimented on the bodies of aboriginal and African peoples. Londa Schiebinger, however, writes that such aboriginal bodies, like those of prisoners, slaves, and the poor, were not "epistemologically weighty." Instead, Lafitau's self-experimentation would have carried more weight with his audience. Londa Schiebinger, "Human Experimentation in the Eighteenth Century: Natural Boundaries and Valid Testing," in *The Moral Authority of Nature*, ed. Lorraine Daston and Fernando Vidal (Chicago: University of Chicago Press, 2004). See also idem, "Medical Experimentation and Race in the Eighteenth-Century Atlantic World," *Social History of Medicine* 26, no. 3 (2013): 364–82.

88. Lafitau, *Mémoire*, 38, 84. Lafitau's *Mémoire* makes several references to Jussieu, suggesting that he and Vaillant were early supporters of his claim to have discovered ginseng and that Jussieu was attempting to grow the plant in the Jardin du Roi in Paris.

89. Jussieu, "Histoire du gin-sem et ses qualités," 1.

90. Ibid., 2. Jussieu also mentioned the possibility that the plant had been first found in Maryland.

91. Ibid.

92. Ibid.

93. Ibid., 3.

94. Jussieu delivered several papers to the Académie in 1718 but, unfortunately, the minutes of the Académie's regular meeting did not record either the title or content of his contributions. It is possible, although it is impossible to verify, that one of these was about ginseng.

95. Geoffroy, for example, sent a letter to Hans Sloane, then secretary of the Royal Society, describing the Académie's ginseng research. See Appleby, "Ginseng and the Royal Society," 130.

96. "Sur le *Gin-seng*," *HMARS* (1718): 51–55. The fact that Bourdelin obtained his samples of ginseng from Jesuits who had recently returned from China was not mentioned.

97. Ibid., 53.

98. Ibid.

99. Ibid.

100. Ibid.

101. Ibid.

102. Ibid.

103. Ibid.

104. Ibid., 54–55.

105. Ibid., 55.

106. On Gaultier, see Lamontagne, "L'influence de Maurepas sur les sciences," 114; Luc Chartrand, Raymond Duchesne, and Yves Gingras, *Histoire des sciences au Québec* (Montréal: Boréal, 1987), 56–60.

107. Gaultier, "Description," 117.

108. Ibid., 117–18.

109. Ibid.

110. Ibid., 118.

111. "Hardi," University of Chicago, ARTFL Dictionnaires d'autrefois, http://artflx .uchicago.edu/cgi-bin/dicos/pubdico1look.pl?strippedhw=hardi.

112. Lafitau, *Mémoire*, 53.

113. Ibid., 64. This was not an unusual concern as knowledge about botanical commodities was published. See Anna Winterbottom, "Producing and Using the 'Historical Relation of Ceylon': Robert Knox, the East India Company and the Royal Society," *British Journal for the History of Science* 42, no. 4 (2009): 524.

114. Block, "New England Merchants," 76.

115. Lozier, "In Each Other's Arms," 314.

116. Ibid., 274.

117. Unlike ginseng, discussions about *capillaire du Canada* (*Adiantum pedatum* or maidenhair fern) explicitly focused on comparison with European varieties. See, for an example, Charlevoix, *Journal d'un voyage*, 4:303. For an example of the collection of *capillaire* by aboriginal peoples, see the account of the practice at Lorette in *JR*, 66:153–55.

118. Lemery, *Traité universel*, 12.

119. The daughters of a prominent Montréal merchant, Marguerite, Marie-Anne, and Marie-Madeleine Desauniers had arrived in the community to establish a trade in clothing and were welcomed both by missionaries who saw in them a means to keep their Iroquoian charges from traveling to Montréal to trade and by Iroquoian people themselves who came to respect their efforts to learn their language as well as their better prices. Louis Lavallée, *La Prairie en Nouvelle-France, 1647–1760: Étude d'histoire sociale* (Kingston: McGill-Queen's University Press, 1992), 235–37. On the activities of the Desauniers sisters and the illicit fur trade more generally, see Jean Lunn, "The Illegal Fur Trade out of New France, 1713–60," *Report of the Annual Meeting of the Canadian Historical Association/La Société historique du Canada* 18, no. 1 (1939): 61–76.

120. Jean-François Lozier argues that "there is little doubt that from early on they were facilitating and encouraging a lively intercolonial trade." Lozier, "In Each Other's Arms," 311.

121. David Preston reminds us that this trip between Kahnawake and Albany "went through a region that remained Mohawk country and a largely Indian landscape." David L. Preston, *The Texture of Contact: European and Indian Settler Communities on the Frontiers of Iroquoia, 1667–1783* (Lincoln: University of Nebraska Press, 2009), 58. See also Lozier, "In Each Other's Arms," 313.

122. Eric Hinderaker, *The Two Hendricks: Unraveling a Mohawk Mystery* (Cambridge, Mass.: Harvard University Press, 2011), 150.

123. Charlevoix, *Journal d'un voyage*, 640.

124. She traded on behalf of the Hôtel-Dieu of Québec, where she was superior for much of the 1730s, 1740s, and 1750s, and received remedies from both Europe (theriac, dried ambergris, epsom salt, etc.) and the wider world (quinquina, jalap, essence of cinnamon, etc.) through the trade. See, for example, "La correspondance de la mère Sainte-Hélène avec Mme Hecquet de La Cloche," *Nova Francia* 4 (1928–29): 123. See also Simon Lorène, "Intérêt pharmaceutique des lettres adressées à l'apothicaire dieppois Féret par les religieuses de l'Hôtel-Dieu de Québec" (PhD diss., Université de Rouen, 2014), 45–48, 126–32.

125. Gaultier, "Description," 117.

126. Lafitau, *Mémoire*, 36–37.

127. Kanna (*Sceletium tortuosum*), the South African plant that Byrd focused on, is not in fact related to ginseng.

128. Appleby, "Ginseng and the Royal Society," 129.

129. Bauer, *The Cultural Geography of Colonial American Literatures*, 189–93. See also William Byrd, *The Dividing Line Histories of William Byrd II of Westover* (Chapel Hill: University of North Carolina Press, 2013), 189. Michael Block explains that it was likely, in fact, his nephew William Beverly who had found the plant. Block, "New England Merchants," 93.

130. Quoted in Marion Stange, *Vital Negotiations: Protecting Settlers' Health in Colonial Louisiana and South Carolina, 1720–1763* (Gottingen: V & R unipress, 2012), 90.

131. "Histoire de l'Académie Royale des Sciences: Année 1718," *Journal des sçavans* 31 (1722): 485.

132. Antoine Furetière, *Dictionnaire universel, contenant généralement tous les mots françois tant vieux que modernes* (The Hague, 1725), vol. 2, n.p.

133. "Quæstio Medica Mane' Discutienda in Scholis Medicorum, die nono Februari 1736," *Journal des sçavans* (1736): 246. See also P. Huard et al., "Une thèse parisienne consacrée au Ginseng en 1736 et présidée par Jean-François Vandermonde," *Bulletin de l'École française d'Extrême-Orient* 60 (1973): 359–74.

134. Du Halde, *Description géographique*, 3:362.

135. Peter Collinson, *"Forget not Mee & My Garden . . .": Selected Letters 1728–1768 of Peter Collinson F.R.S.* (Philadelphia: American Philosophical Society, 2002), 70.

136. See, for example, "Dissertation sur la célèbre terre de Kamtschatka," *Mémoires pour l'histoire des sciences & des beaux arts* (1737): 1159; Jacques Savary des Bruslons and Philemon-Louis Savary, *Dictionnaire universel de commerce contenant tout ce qui concerne le commerce* (Geneva: Cramer & Philibert, 1750), 2:651; Nicolas Lemery, *Dictionnaire universel des drogues simples* (Paris: Chez Theophile Barrois, 1759), 127.

137. "GIN-SENG," *Encyclopédie, ou dictionnaire raisonné des sciences, des arts et des métiers*, ed. Denis Diderot and Jean le Rond D'Alembert, University of Chicago: ARTFL Encyclopédie Projet (Winter 2008 edition), ed. Robert Morrissey, http://encyclopedie.uchicago.edu/.

138. Ibid.

139. M. Geoffroy, "Moyen de préparer quelques racines à la manière des orientaux," *HMARS* (1740): 96–99.

140. Jacques Savary des Bruslons and Philemon-Louis Savary, *Dictionnaire universel de commerce contenant tout ce qui concerne le commerce* (Geneva: Cramer & Philibert, 1742), 2:627.

141. Ibid.

142. Kalm, *Voyage de Pehr Kalm*, 224–25.

143. Quoted in Evans, "Ginseng: Root of Chinese-Canadian Relations," 13.

144. Quoted in Preston, *The Texture of Contact*, 211.

145. Gail D. MacLeitch, "'Red' Labor: Iroquois Participation in the Atlantic Economy," *Labor* 1, no. 4 (2004): 81.

146. Louis Franquet, *Voyages et mémoires sur le Canada par Franquet* (Montréal: Editions Elysée Montréal, Institut Canadien de Québec, 1974), 99.

147. Kelly Yvonne Hopkins, "A New Landscape: Changing Iroquois Settlement Patterns, Subsistence Strategies, and Environmental Use" (PhD diss., University of California, Davis, 2010), 135.

148. Evans, "Ginseng: Root of Chinese-Canadian Relations," 15; Gail D. MacLeitch, *Imperial Entanglements: Iroquois Change and Persistence on the Frontiers of Empire* (Philadelphia: University of Pennsylvania Press, 2011), 92; Timothy J. Shannon, "Dressing for Success on the Mohawk Frontier: Hendrick, William Johnson, and the Indian Fashion," *William and Mary Quarterly* 53, no. 1 (1996): 19–20.

149. "Lettre de Duquesne et Bigot au ministre," October 15, 1752, C11A, 98:8–11v, ANOM.

150. Franquet, *Voyages et mémoires*, 60.

151. "Ordonnance qui fait défense à tous habitants . . . de cueillir du ginseng," A6, 20:105–7, ANOM.

152. Lafitau, *Mémoire*, 67; Byrd, *The Dividing Line Histories*, 189.

153. McGraw et al., "Ecology and Conservation of Ginseng (Panax quinquefolius) in a Changing World," 62.

154. Lafitau, *Mémoire*, 74–79.

155. Ibid., 67.

156. Lafitau, *Mémoire*, 14.

157. This quote is from a transcription of "Lettre de Pierre Poivre au Comité secret à Canton, le 31 décembre 1750," *Pierre Poivre & Compagnie*, http://www.pierre-poivre.fr/doc-50-12-31a.pdf.

158. Evans, "Ginseng: Root of Chinese-Canadian Relations," 11–12.

159. Ibid., 12. By the 1750s officials began to complain about improperly processed, stored, and collected plants. By 1752 these same officials lamented the fall in prices that resulted from decreased interest in Canadian ginseng in the Chinese market. See, for a representative example, "Omissions à suppléer dans le mémoire," 1752, C11A, 98:460–61, ANOM.

160. Rénald Lessard, "Pratique et praticiens en contexte colonial: Le corps médical canadien aux 17e et 18e siécles" (PhD diss., Université Laval, 1994), 174.

161. Guillaume-Thomas Raynal, *Épices et produits coloniaux* (Paris: Ed. la Bibliothèque, 1992 [1770]), 65.

162. Lafitau, *Mémoire*, 68.

163. Evans, "Ginseng: Root of Chinese-Canadian Relations," 14.

CONCLUSION

1. Louis-Antoine de Bougainville, *Ecrits sur le Canada: Mémoires, journal, lettres* (Sillery: Septentrion, 2003), 79.

2. Ibid.

3. Ibid.

4. See, for recent work that has emphasized the tensions within enlightenment thought, David Allen Harvey, *The French Enlightenment and Its Others: The Mandarin, the Savage, and the Invention of the Human Sciences* (New York: Palgrave Macmillan, 2012); Muthu, *Enlightenment Against Empire*.

5. Bougainville, *Ecrits sur le Canada*, 80.

6. Kalm, *Voyage de Pehr Kalm*, 3. See also Chapter 5.

7. On the (frequently complicated) appeal of plantation landscapes in the eighteenth century, see Casid, *Sowing Empire*; Britt Rusert, "Plantation Ecologies: The Experimental Plantation in and Against James Grainger's *The Sugar-Cane*," *Early American Studies* 13, no. 2 (2015): 341–73. On the debate over whether to keep Canada or Guadeloupe in the negotiations at the end of the Seven Years' War, see Helen Dewar, "Canada or Guadeloupe?: French and British Perceptions of Empire, 1760–1763," *Canadian Historical Review* 91, no. 4 (2010): 637–60.

8. Harvey, *The French Enlightenment and Its Others*, 1.

9. Curran, *The Anatomy of Blackness*, 11. The literature on the emergence of race in the eighteenth century is both enormous and beyond the scope of this conclusion. See, for examples, Nicholas Hudson, "From 'Nation' to 'Race': The Origin of Racial Classification in Eighteenth-Century Thought," *Eighteenth-Century Studies* 29, no. 3 (1996): 247–64; Londa L. Schiebinger, *Nature's Body: Gender in the Making of Modern Science* (New Brunswick, N.J.: Rutgers University Press, 2004); Roxann Wheeler, *The Complexion of Race: Categories of Difference in Eighteenth-Century British Culture* (Philadelphia: University of Pennsylvania Press, 2000).

10. Thierry Hoquet, "Biologization of Race and Racialization of the Human: Bernier, Buffon, Linnaeus," in *The Invention of Race: Scientific and Popular Representations*, ed. Nicolas Bancel, Thomas David, and Dominic Thomas (New York: Routledge, 2014). On earlier formulations of race, see Aubert, "'The Blood of France.'" For the interaction between these various "fields," see Claude-Olivier Doron, "Race and Genealogy: Buffon and the Formation of the Concept of 'Race,'" *Humana Mente*, no. 22 (2012): 75–110.

11. Claude Blanckaert, "Of Monstrous Métis? Hybridity, Fear of Miscegenation, and Patriotism from Buffon to Paul Broca," in *The Color of Liberty: Histories of Race in France*, ed. Sue Peabody and Tyler Stovall (Durham, N.C.: Duke University Press, 2003), 44. See also John H. Eddy, "Buffon's *Histoire naturelle*: History? A Critique of Recent Interpretations," *Isis* 85, no. 4 (1994): 651–52.

12. The best overview of this argument remains Gerbi, *The Dispute of the New World*, ch. 1.

13. Cañizares-Esguerra, *How to Write the History of the New World*, 46–47.

14. This determinism was challenged on several fronts. See, for discussions of disputes around Buffon's claims, Gerbi, *The Dispute of the New World*. Doron suggests that assumptions about manipulation were foundational to Buffon's natural history, owing to debts to the science of horse breeding. Doron, "Race and Genealogy," 80. For analysis of plans to "engineer" races in the French Caribbean, see William Max Nelson, "Making Men: Enlightenment Ideas of Racial Engineering," *American Historical Review* 115, no. 5 (2010): 1364–94.

15. While avoiding the determinism of Buffon, science studies scholars have been similarly interested in reintroducing ecological relationships into the study of social relationships. See, for an example, Linda Lorraine Nash, *Inescapable Ecologies: A History of Environment, Disease, and Knowledge* (Berkeley: University of California Press, 2006).

16. On this episode, see Dugatkin, *Mr. Jefferson and the Giant Moose*.

17. "I love peace": quoted in Dewar, "Canada or Guadeloupe?" 638; "a few acres of snow": quoted in Glen Norcliffe and Paul Simpson-Housley, "No Vacant Eden," in *A Few Acres of Snow: Literary and Artistic Images of Canada*, ed. Glen Norcliffe and Paul Simpson-Housley (Toronto: Dundurn Press, 1992), 1.

18. Gilbert Chinard, "Influence des récits de voyages sur la philosophie de J. J. Rousseau," *PMLA* 26, no. 3 (1911): 476–95; Wåhlberg Martin, "Littérature de voyage et savoir: La méthode de lecture de Buffon," *Dix-huitième siècle* 42, no. 1 (2010): 599–616; Safier, *Measuring the New World*, ch. 6.

19. Ouellet and Beaulieu, "Avant-propos," 181.

20. See Patrick Wolfe, *Settler Colonialism and the Transformation of Anthropology: The Politics and Poetics of an Ethnographic Event* (London: Cassell, 1999); Veracini, *Settler Colonialism.*

21. Veracini, *Settler Colonialism*, 33.

22. Lestringant, "Champlain, Lescarbot et la 'conférence' des histoires," 336; Zecher, "Marc Lescarbot."

23. On the epic narratives of the Jesuits, see Pioffet, *La tentation*. For a discussion of Sagard's evocation of a providential design, see Warwick, introduction to *Le grand voyage*, 49.

24. See Chapter 4.

25. Patrick Wolfe, "Settler Colonialism and the Elimination of the Native," *Journal of Genocide Research* 8, no. 4 (2006): 388.

26. On this sort of disavowal, see Veracini, *Settler Colonialism*, ch. 3.

27. On the Fox Wars, see Brett Rushforth, "Slavery, the Fox Wars, and the Limits of Alliance," *William and Mary Quarterly* 63, no. 1 (2006): 53–80; White, *The Middle Ground*, ch. 4. Benjamin Madley has recently identified the Fox Wars as genocidal in "Reexamining the American Genocide Debate: Meaning, Historiography, and New Methods," *American Historical Review* 120, no. 1 (2015): 113.

28. Mario Blaser and Arturo Escobar, "Political Ecology," in *Keywords for Environmental Studies* (New York: New York University Press, 2016), 164.

29. Lisa Slater, "'Wild Rivers, Wild Ideas': Emerging Political Ecologies of Cape York Wild Rivers," *Environment and Planning D: Society and Space* 31, no. 5 (2013): 769.

30. David Rich Lewis, *Neither Wolf nor Dog: American Indians, Environment, and Agrarian Change* (Oxford: Oxford University Press, 1994), 7. On "peasant" farming policies, see Sarah Carter, *Lost Harvests: Prairie Indian Reserve Farmers and Government Policy* (Kingston: McGill-Queen's University Press, 1990).

31. Joshua L. Reid, *The Sea Is My Country: The Maritime World of the Makahs, an Indigenous Borderlands People* (New Haven, Conn.: Yale University Press, 2015). See, for a discussion of the genetic effects of commercial whaling in this region, S. Elizabeth Alter, Seth D. Newsome, and Stephen R. Palumbi, "Pre-Whaling Genetic Diversity and Population Ecology in Eastern Pacific Gray Whales: Insights from Ancient DNA and Stable Isotopes," *PLoS ONE* 7, no. 5 (2012): 35039.

32. Marsha L. Weisiger, *Dreaming of Sheep in Navajo Country* (Seattle: University of Washington Press, 2009); Mark David Spence, *Dispossessing the Wilderness: Indian Removal and the Making of the National Parks* (Oxford: Oxford University Press, 1999).

33. Winona LaDuke, *Recovering the Sacred: The Power of Naming and Claiming* (Cambridge, Mass.: South End Press, 2005), 167–90. The literature on biopiracy is vast. For introductions, see Jack Ralph Kloppenburg, *First the Seed: The Political Economy of Plant Biotechnology, 1492–2000*, 2nd ed. (Madison: University of Wisconsin Press, 2004); Thom van

Dooren, "Inventing Seed: The Nature(s) of Intellectual Property in Plants," *Environment and Planning D: Society and Space* 26, no. 4 (2008): 676–97.

34. Geniusz, *Our Knowledge Is Not Primitive*; Robin Wall Kimmerer, *Braiding Sweetgrass: Indigenous Wisdom, Scientific Knowledge and the Teachings of Plants* (Minneapolis: Milkweed Editions, 2013); Gary Paul Nabhan, *Enduring Seeds: Native American Agriculture and Wild Plant Conservation* (Tucson: University of Arizona Press, 2002); Nancy Turner, *Ancient Pathways, Ancestral Knowledge: Ethnobotany and Ecological Wisdom of Indigenous Peoples of Northwestern North America* (Kingston: McGill-Queen's University Press, 2014).

INDEX

ACKNOWLEDGMENTS

I've now spent over a decade thinking and writing about intellectual networks and the strange alchemy that produces good, collaborative research. I can read almost each line of *A Not-So-New World* and see how conversations with friends, colleagues, and mentors have shaped my thinking and helped me usher this project from conception through to the book that it is today. I've been blessed to be able to join and pass through a dizzying array of communities along the way who have each left their mark on this text. I am grateful to them all.

I owe my greatest thanks to those who have taken the time to read and offer feedback on various iterations of this book that I've written over the years. Allan Greer both introduced me to New France and helped me understand how to frame this project in broader American and Atlantic contexts. Heidi Bohaker, Nicholas Dew, Kenneth Mills, and Natalie Zemon Davis offered insightful comments on a first draft of the project that helped me immensely. At the McNeil Center for Early American Studies and the University of Pennsylvania Press, both Daniel Richter and Robert Lockhart pushed me to think in broader terms still and worked patiently through the manuscript with me. I am particularly grateful to Dan for organizing a manuscript workshop and to Roger Chartier, Projit Mukharji, Susan Scott Parrish, Lisa Rosner, and Daniel Usner for their contributions. I'd also like to thank the anonymous reviewers of this manuscript and the staff at the University of Pennsylvania Press, whose perceptive critiques helped me immensely.

I'm grateful to the organizers and audiences of seminars and conferences that provided important forums for untangling methodological knots and charting richer courses in my research. These include the Early Modern Studies Institute–William & Mary Quarterly workshop, the Columbia Seminar on Early American History and Culture, "Explorations, Encounters, and the Circulation of Knowledge, 1600–1830" at the Clark Library, the Omohundro Institute for Early American History and Culture colloquium, the McNeil

Center for Early American Studies seminar, the Center for Early Modern History at the University of Minnesota, "American Oecologies" at the John Carter Brown Library, "De l'observation à l'inscription: Les savoirs sur l'Amérique entre 1600 et 1830 dans les textes d'expression française" in Montréal, the Summer Academy of Atlantic history, two Jesuit Studies workshops at Northwestern University, and the 2009 Harvard Atlantic History Seminar. This project was transformed by the opportunity to share it with friends and colleagues at these events, and I thank the organizers, audiences, and other participants alike.

At the University of Toronto, I couldn't have asked for better friends and colleagues than my fellow early Canadianists Helen Dewar and Jean-François Lozier. Toronto offered exciting opportunities to be disciplinarily promiscuous, and I thank Paul Cohen, Heather Coiner, Victoria Freeman, Lisa Helps, Adrienne Hood, Michelle Murphy, Jan Noel, Steve Penfold, Ian Radforth, Sheena Sommers, and Olivia Wilkins for sharing their expertise with me at pivotal moments. As a visiting scholar at Dartmouth, I found a welcoming and enriching community in the Native American studies program and insightful readers in Colin Calloway, Bruce Duthu, Vera Palmer, Melanie Benson Taylor, and Dale Turner. At the McNeil Center for Early American Studies, I was able to join two great cohorts of fellows, and I'm grateful to Susan Brandt, Sarah Chesney, Andrew Fagal, Hannah Farber, Mitch Fraas, Cassandra Good, Glenda Goodman, Rachel Herrmann, Craig Hollander, Kathleen Howard, Matt Karp, Adam Lewis, Nenette Luarca-Shoaf, Whitney Martinko, Mark Mattes, Dael Norwood, Seth Perry, Edward Pompeian, Ben Reed, Jessica Roney, Sarah Scheutze, Samantha Seeley, Danielle Skeehan, Steven Smith, and Tristan Tomlinson for providing the encouragement, conversation, and community in which this book took shape. Northeastern University has been a good home to me these past few years, and I have been privileged to find an amazing group of friends and colleagues in Victoria Cain, Gretchen Heefner, Benjamin Schmidt, and Philip Thai. My other colleagues, including Tim Cresswell, Bill Fowler, Laura Frader, Laura Green, Tom Havens, Elizabeth Maddock Dillon, Tony Penna, Harlow Robinson, and Louise Walker, deserve thanks for their warm welcome and generous intellectual engagement with my work.

I've been tremendously lucky in meeting rigorous scholars who also understand the value of conversation and collaboration. Kathleen Murphy and Cameron Strang were both fantastic interlocutors and wonderful coauthors. Coll Thrush was generous with his time and his expertise as I worked on a conclusion that made this the book I wanted it to be. Robert Morrissey and Benjamin Madley have consistently been fantastic friends and generous readers. Andreas Motsch has been an inspiring collaborator. Others have read drafts, talked through challenges, created opportunities to share ideas and conversation, and, in short, been wonderful support. These include Guillaume Aubert, Allison Bigelow, John Bishop, Eva Botella-Ordinas, Ben Breen, Celine Carayon, Joyce Chaplin, Matthew Crawford, Joseph Cullon, Catherine Desbarats, Victoria Dickenson, Robert Englebert, Scott Ferguson, Darcie Fontaine, Claire Gherini, Gilles Havard, Florence Hsia, Kristin Huffine, Julia Irwin, Drew Lipman, Peter Mancall, Bertie Mandelblatt, Ben Marsh, Michelle Molina, Vanessa Mongey, Bruce Moran, Eric Otremba, Thomas Peace, Steve Prince, François Regourd, Michael Reidy, James Rice, Dan Rück, Brett Rushforth, Amy Rust, Neil Safier, Gordon Sayre, Susan Sleeper-Smith, Andrew Sturtevant, Guillaume Teasdale, Fredrika J. Teute, Micah True, Karin Velez, Aaron Walker, Molly Warsh, Sophie White, Thomas Wien, Kelly Wisecup, Karin Wulf, and Anya Zilberstein. I must also thank librarians and staff at the Bibliothèque nationale de France, Renald Lessard at the Bibliothèque at Archives nationales de Québec, Cécile Aupic at the Muséum national d'Histoire naturelle, Timothy Dickinson at the Royal Ontario Museum, Ruth Newell at the Harriet Irving Botanical Gardens at Acadia University, and Roy Goodman at the American Philosophical Society.

I also owe a great deal to several governmental agencies, institutions, and libraries that saw the merits of this work—in many cases even before I did—and made travel to libraries, archives, and conferences possible through their financial support. The generous support of the Social Sciences and Humanities Research Council of Canada, the Ontario Ministry of Training, Colleges and Universities, Associated Medical Services, Inc., and the University of Toronto allowed me time to focus on research and writing. A fellowship from the McNeil Center for Early American Studies provided invaluable time to rework this project and do new research. Northeastern University has provided both leave and funding to finish research for this project. I am also tremendously grateful to the John Carter Brown Library, the Newberry Library, and the American Philosophical Society both for their financial support and, more importantly, for introducing me to many colleagues who have shaped my work profoundly and who, since my trips to these libraries, have in many cases become close friends.

Finally, I want to thank friends and family for their part in the production of this book. I couldn't ask for better cheerleaders than my parents and stepparents and Nora, Allison, and Robin. In addition, I've been particularly fortunate in finding a partner, Sari Altschuler, who has also proven to be an incomparable collaborator, editor, and occasional coauthor. Sari's fingerprints are all over this book, and it's a delight to read through and—in a sort of intellectual Where's Waldo?—see hints of conversations that began at the McNeil Center for Early American Studies and that have continued over dinners, at conferences, in the woods, and on mountains. *A Not-So-New World* is only one small piece of a partnership that deepens with each passing year.

* * *

Portions of this book have been published as articles. Chapter 1 is a revised version of "Wildness Without Wilderness: Biogeography and Empire in French North America," *Environmental History* 22, no. 4 (2017): 643–67. Portions of Chapter 6 were published as "The Natural History of Colonial Science: Joseph-François Lafitau's Discovery of Ginseng and Its Afterlives," *William & Mary Quarterly* 73, no. 1 (2016): 37–72. I thank the editors of these journals for permission to reprint them here and for their help with this project.